SCIENCE AND POLITICS OF FOREIGN AID

Science and Politics of Foreign Aid

Swedish Environmental Support to the Baltic States

by

Björn Hassler
Södertörn University,
Huddinge, Sweden

SPRINGER-SCIENCE+BUSINESS MEDIA, B.V.

A C.I.P. Catalogue record for this book is available from the Library of Congress.

ISBN 978-94-010-3969-7 **ISBN 978-94-010-0123-6 (eBook)**
DOI 10.1007/978-94-010-0123-6

Printed on acid-free paper

TABLE OF CONTENTS

PREFACE

The increasing interdependence between nations in the modern world is one of the most discussed issues in contemporary studies of international relations. The interaction between countries, corporations, organizations, and people across country borders tends to increase wealth and a multitude of various personal and collective benefits, as well as creating friction and conflicts of interests. This book concerns the potential conflict between national self-interest and morality, in terms of provision of foreign aid. More specifically, it concerns the Swedish environmental support to the Baltic States during the first half of the 1990s. This particular support is of extraordinary interest for several reasons. Not only is it an example of the difficulties in keeping trade and aid issues apart. It is also a case where the gains at stake are tangible for both the recipient and donor countries. The trans-boundary nature of some of the most important environmental issues of our time makes the fruits from such cooperation important to the donor as well. Finally, the rapid transformation of the Baltic States from Moscow-controlled satellites to modern liberal democracies makes the study of the potential for joint gains from cooperation in this region politically important, as well as scientifically tantalizing.

When approaching the most important issue in this study – the rationales behind the Swedish environmental assistance to the Baltic States - it became increasingly clear to me that in order to produce a scientifically proficient study in social science it would be necessary to combine theoretical rigor with empirical depth. The ease with which one could fall into the mire of empirical bookkeeping that had little bearing on the production of scientific knowledge, was as great and apparent as the temptation to endlessly dwell on theoretical puzzle solving. Therefore, my ambition has been to find a middle-road, where theory is used as a sharp instrument to discern reality. Without theoretical rigor, the work would have been extremely cumbersome and probably less successful than what has been achieved. Without attempts to explain real-world events, nothing of scientific interest would have been accomplished. This ambition did, however, require a significant personal effort, as well as help and advice from many colleagues. Fortunately, I have not lacked assistance from a large number of people.

This book has, to a large extent, been based on the findings of my PhD studies at the Department of Water and Environmental Studies (1995-2000), at the University of Linköping in Sweden. The person to whom I feel most deeply indebted is Professor Elfar Loftsson, my advisor. Since my arrival in the department, he has always been a very good friend. Numerous are the times we have discussed issues concerning, for example, science and politics over a cup of coffee or a glass of beer. Such conversations always left me with a sense of increased commitment and a fervent desire to continue my work. Elfar's thorough reading of my manuscripts at countless stages of completion has been of the utmost importance to me. His sharp observations of inconsistent arguments and empirical fallacies have forced me to reconsider previous lines of thought, and thus increased the quality of the study.

This book has been written within the research project *Social Change and Democracy in the Baltic Sea Region*. I feel very grateful to the other members of this research group for their support as well as for their critical remarks. In addition to Elfar Loftsson, who has headed this research project, Wojciech Szrubka and Yonhyok Choe have made significant contributions to my work, and have also ensured my social well-being.

During the work on this book, I have had the pleasure of being employed by Södertörns högskola. Being a member of the large political science team at Södertörn has been an inspiration. From the first day, I felt very welcome. It is a young University/College where everyone has been excited about the prospect of building a new academic center. I am thankful that I had the opportunity to be a part of this important endeavor.

A number of external professionals have been of great importance to my work, as readers of early parts of the manuscript. My brother John made several suggestions for improvements in the theoretical parts of this study. His sharp observations from an economist's perspective have been most valuable. Additional insightful comments from various disciplinary perspectives have been given to me by Jan Lindvall and Örjan Widegren at the University of Linköping, Kenneth Kollman at the University of Michigan, and Lasse Ringius at the UNEP Collaborating Centre on Energy and Environment. I would also like to give a special thank you to Åsa Danielsson and Charlotta Pers for their contribution to the development of my skills in statistical methods. The anonymous reviewer assigned by Kluwer Academic Publishers has provided me with numerous valuable comments and suggestions on how the text could be improved, for which I am deeply grateful. Working in cooperation with the staff at Kluwer Academic Publishers in general, and Mrs. Gloria Verhey in particular, has been a pleasure. The encouraging and constructive dialogue during the process of writing has been valuable to me. The kind and helpful assistance given to me when making the final preparation of the manuscript has been especially useful.

Finally, I would like to thank the administrative staff of both the Department of Water and Environmental Studies and Södertörns högskola for their kind and professional assistance. This research study was originally financed by the Department of Water and Environmental Studies and the Baltic Sea Fund (Östersjöstiftelsen). I am very grateful to them both for their support.

CHAPTER 1

INTRODUCTION

1. FROM REALISM TO ENVIRONMENTAL INTERDEPENDENCE

The Second World War shaped the modern study of international relations in several important ways. The destruction and evil demonstrated by the war convinced most scholars that the idealistic approach was not in accordance with reality. *Realism* was essential in order to find a more accurate description of the interaction between sovereign states in a world without institutions above the state level with coercive power. The outcome of the interaction was determined not so much by ideological or moral standpoints as by the power of individual states or coalitions. The development in international relations studies in the 1950s seemed to confirm this conviction. The need to understand the processes in the international system as a struggle between sovereign states was epitomized for example by Hans J. Morgenthau in his seminal work *Politics among nations: the struggle for power and peace*, first published in 1948.

During the 1950s, the international scene became increasingly dominated by the tension between East and West, between the Warsaw Pact and NATO. Scholars in international relations described this dichotomization as the emergence of a bipolar international system. Kenneth Waltz emphasized the important distinction between a bipolar and a multipolar world system. The issue was still about power and military capability. According to Waltz' analysis in his *Theory of international relations*, published in 1979, a bipolar structure was preferable to a multipolar structure, since the power balance would be more stable in the former.

The occurrence of periods of détente between East and West as well as the increasing importance attached to the interdependent nature of the relations between nations in the late 1970s and early 1980s, indicated that the classical power-approach in studies of international relations could only partially explain observed behavior. Although military capability was still very important, it was argued, by for example Robert Keohane (1977) and Robert Gilpin (1981), that the *incentives* for choosing different strategies had to be added to the capability parameter. It was not reasonable to assume that states would act as expansionistically as their military capability allowed them, without regard to the cost of such policies. In the age of nuclear devices, it was also appreciated that even if a state possessed superior strength, this superiority could not easily be converted into influence over other states' behavior.

Instead of relying only on military capability as the explanatory variable, a more reasonable assumption seemed to be that states choose strategies from expected cost-benefit ratios. States choose strategies to increase the welfare of dominant domestic interests. Not necessarily because of the benevolence of the governments, but rather due to the power and preferences of national interest groups. It seemed to become more and more obvious that the interest of the states was not confined to issues of war and peace. As the volume of trade and mobility of capital kept increasing, the economic interdependence in the international system became a progressively imperative feature. Larger trade volumes meant increased prosperity and thereby amplified economic power.

In the late 1970s and early 1980s, the transnational nature of many kinds of environmental issues became apparent to scientists as well as to electorates. The United Nations Environment Programme (UNEP) had been created as a direct outcome of the 1972 Stockholm Conference on Human Environment. In the 1980s, the World Bank (IBRD) began to change its policies in a more environmentally adapted direction, due to criticism from several of its member countries and Non-Governmental Organizations (NGOs). Slowly, these changes began to change practices in the field. At the regional level, several conventions were formulated at about the same time as UNEP was created, for example the *Helsinki Convention on the Protection of the Marine Environment of the Baltic Sea Area.* Increasing resources were allocated to studies of environmental disturbances crossing national borders in most Western countries. Negotiations targeted towards joint action to tackle transnational environmental issues are seldom a zero sum game. There is almost always a potential for Pareto-sanctioned improvements, where at least one participant gains without others losing. This does not mean, however, that the benefits will be equally distributed, or that there are no incentives to free-ride, to let others bear the costs while benefiting from their action. On the contrary, typically the action taken will benefit the actors unevenly and there will be serious incentives to free-ride. Both factors tend to induce strategic behavior that might cause inertia, despite the potential for joint gains.

The classical approach emphasizing power and military capability as the main explanatory variable has proved to be unsuitable for the analysis of state interaction on environmental issues. Brute force is seldom a feasible instrument, thus making such threats less credible. Interest has instead been focused on two different paths propounding the explanation of successful, as well as unsuccessful, initiatives to joint action. The first is based on the incentive approach. To explain outcomes in, for example, transnational environmental negotiations the interests of the involved actors have to be analyzed. Based on these interests, hypotheses about preferred strategies of the actors can be inferred. The other broad category does not focus on the incentives facing the actors, but rather on the potential to reach a cognitive mutual understanding of the problem at hand. It is a priori assumed that there exists a solution acceptable to all actors. If such an understanding of an acceptable solution can be reached, it is assumed that the actors will work towards this solution.

In this study, the Swedish environmental assistance to the Baltic States in the period 1991-1996 is analyzed, using the approach based on the interests of the actors. This assistance will be interpreted as a mechanism to alter the behavior of the Baltic States, when the use of force is neither available, nor desirable. It is assumed that the transfer of Swedish resources is used to change the preferences of the Baltic States over potential environmental investments, in directions desirable to Sweden. When Sweden is covering some of the investment costs in particular environmental sectors, the governments of the Baltic States are induced to cover the remaining costs in order to get the investments implemented. It should be emphasized, however, that since the realization of a particular cooperative project requires consent from both the donor and the recipient, the outcome is expected to be mutually beneficial. In other words, the "game" being played is not zero sum, but one with a positive overall sum.

2. ALTRUISM AND SELF-INTEREST: THE AIM OF THE STUDY

As a starting point it seems reasonable to assume that the Swedish environmental support to the Baltic States is perceived as being part of general foreign assistance and thus primarily guided by an intention to help people suffering from poverty, oppression, diseases, environmental hazards and similar hardships. The Swedish assistance to Central and Eastern Europe (CEE) was, in the first years, financed from the aid budget, and the need to support the countries in transition in their struggle to become de facto sovereign states was expressed time and again in government propositions and declarations, as well as in parliamentary bills by the political parties in opposition. Although there were certainly also other less altruistic aspects influencing Swedish foreign aid in general, the official ambition as well as the perception of the public was that foreign aid should primarily benefit poor people in poor countries. From this perspective, the apparent aspects of self-interested considerations in the Swedish environmental assistance to the Baltic States seem surprising. This observation forms the basis of this study, where the primary objective is to analyze to what extent this particular assistance can be explained by self-interested Swedish behavior. It should be emphasized that self-interested behavior does not necessarily mean that the interests of the recipient country are compromised. A self-interested actor finds no pleasure in obstructing the policies of other actors per se. Self-interest here means that the interests of the other actors are not taken into consideration. But if it is, for example, the case that mutual gains might be realized, a self-interested actor would have no problems at all with the fact the others also gain from a particular initiative.

Foreign aid is defined as purely altruistic if the *only* determinant of relevance to the donor country is its perception of the recipient's utility from the assistance undertaken. The donor country in this case only cares about the recipient's gains. Domestic interests are only addressed if this does not negatively affect the utility derived from the assistance to the recipient. This does *not* imply, however, that efficiency issues are irrelevant. The common notion of cost-benefit ratio may still be used, but *only* the utility of the recipient and *only* the cost for the donor country would be relevant. Trying to maximize recipient utility to a given cost does not violate the definition of altruism above.

In contrast to this truly philanthropic donor, the ideally self-interested one *only* cares about the effect on domestic considerations. It is difficult, indeed, to imagine any aid policies at all under such auspices, since the money spent on assistance would almost always be of more utility to the donor not taking the path through aid disbursements. Why not allocate the resources directly into the domestic economy, and transform the aid bureaucracy to e.g. an industry promotion agency? The self-interested actor might not do this for two reasons. First, when we talk about an ideally self-interested actor this refers to a theoretical construction, used as a point of reference. Lack of empirical cases corresponding to this ideal type, does not depreciate its validity as a conceptual benchmark. Second, an actor trying to maximize self-interest never does so without restrictions. In most cases, the restrictions on maximizing self-interest are economic, such as for example the budget restriction. Here, however, political restraints are more relevant. Restraints, which may at first glance seem to be incompatible with self-interest, but following closer analysis might in fact be consistent with this assumption. It may not, for example, be politically possible to use all Swedish assistance resources for procurement of goods and services made in Sweden by Swedish firms, even if this would benefit Sweden more than using some of the resources to, for example, support United Nations activities. What gives the highest, marginal utility may not be all that clear in all cases. What is important, however, is that we do not want to incorporate altruistic assumptions into the self-interested utility function. To do this would deprive us of analytical stringency, since any type of behavior in that case could be compatible with utility maximizing behavior.

In the world of empirical observations, behavior is probably never fully compatible with either ideal model, neither perfect altruism, nor complete self-interest. Swedish development assistance is not, and has probably never been, driven by purely altruistic considerations or based exclusively on national self-interest. The point here is that our assumption states that self-interest is the *most important* factor deciding state behavior. This was succinctly put by K. R. Monroe in her review of the book *Pathologies of rational choice: A critique of applications in political science* by D. P. Green & I Shapiro (1994): "Recognizing that there are important qualifications and exceptions to self-interest does not mean it is not a central tendency, any more than the fact that some parents do not love their children does not lead us to reject as false the premise that most parents *do* love their children and to construct our theories of parental behavior on that assumption." (Monroe 1997) If it can be shown that self-interest was an important factor determining observed behavior, our endeavor has been fulfilled. The ambition is not to present *all* factors influencing the states choices of strategies. The motivation to use a model based on the assumption of self-interested actors is thus pragmatic. We do believe that the assistance can be better explained assuming self-interested rather than altruistic behavior. It is important to recognize that this is not an ontological issue, but an epistemological one, where the goal is to find the most productive approach. The fruitfulness of this approach will then be possible to judge after having analyzed the empirical data.

To a large extent the international environmental discourse has in the last decade to a large extent been focused on "global issues", "common responsibilities" etc. To a certain extent, the use of such expressions has had political undertones. There have been deliberate attempts from individual states as well as Inter Governmental Organizations (IGOs) to change the behavior of other actors in certain directions. Because of analytical reasons, no assumptions are made about the existence of some kind of a common environmental strategy in this study. The only strategies assumed are those of the individual states. This does not mean, however, that several states may not have similar issues on the priority list of their national agendas. If an environmental initiative is financed with domestic resources, the outcome may or may not affect other countries, but the relevant factor is the relation between cost and domestic utility. If some other country, on the other hand, implements an environmental project that affects countries other than itself, this utility enters the decision-situation of those other countries. This opens up the possibility for strategic behavior. In order to achieve maximum domestic utility, it would be an advantage if other countries could be persuaded to choose projects that had beneficial effects for us. It is in this situation that environmental assistance can be used as an instrument for enlarging domestic utility. This possibility is analyzed in this study.

3. THE ENVIRONMENTAL ISSUES

There are two broad categories of environmental issues that have particular salience in this study due to their great potential to affect more than one country: Emissions to the air and emissions into the Baltic Sea. The emissions of NO_X, SO_X and CO_X to the air all become dispersed over very large geographical areas. However, due to prevailing average wind conditions, Sweden is generally less affected by such emissions from the Baltic States, than what otherwise might be expected (and consequently comparably more affected by emissions from e.g. the UK). Despite this fact, Sweden has been very active in this area and supported several comparably large projects in the Baltic States primarily aimed at reducing emissions of so-called greenhouse gases. Reductions of these kinds of gases have global implications, although the consequences for individual countries vary considerably. Despite the fact that the Nordic contribution of carbon dioxide is only about one percent of the global output, Sweden has kept a quite high profile in the international discussions on the reduction of greenhouse gases (Nordic Council of Ministers 1996). In 1993 the target was set to stabilize the emissions of carbon dioxide at the 1990 level before the year 2000, and to subsequently reduce emissions. As part of the Swedish climate strategy, considerable amounts have been invested in Estonia in particular, on the grounds that investments there were more cost-efficient than in Sweden.

Emission of radioactive substances into the air has been a different matter. The nuclear power plant reactor in Ignalina is similar to the one in Chernobyl, where a serious accident occurred in 1986. Although it is very uncertain, due to specific meteorological conditions at the time of the emission, what regions would be affected from another fall-out, the Chernobyl accident clearly showed that Sweden might be severely affected.

The reduction of discharges of harmful substances and nutrients into the Baltic Sea has been the most important target for Swedish environmental assistance to the Baltic States. As late as the 1950s, the Baltic Sea was a comparably clean sea. Large-scale industrial activities were at that time relatively rare in the surrounding countries, the number of cars comparably few, and the agro-forestry sector used considerably less intensive production methods (i.e. smaller volumes of fertilizers and pesticides). Today, this is no longer the case.

> Pollution now threatens the entire Baltic Sea catchment area as well as the Baltic Sea itself, and ultimately the health and well-being of the 80 million people who live there (HELCOM 1993a: 6).

The biological equilibrium of the Baltic Sea is extraordinarily sensitive due to several natural factors. The low salinity of the brackish water makes many organisms more susceptible to changes in environment, since their natural habitat is often in either the sea or in inland lakes. In addition, vertical variations in salinity make the water stratified, thus lowering the transportation of oxygen. This is one reason behind the quite widespread occurrence of very low concentrations of oxygen and almost life-less seabeds. Another factor making the Baltic Sea especially vulnerable to the deposition of harmful substances is the long residence period (25-40 years) of its water volume (HELCOM 1993b: 154). Due to the narrow and shallow inflow through the Belts and the Sound, North Sea water replenishes the Baltic Sea water very slowly. The largest inflows come from freshwater sources, which often carry considerable amounts of pollutants (Stålnacke 1996).

The most important environmental problem today related to the Baltic Sea is probably the high load of nutrients, which are causing increased eutrophication (HELCOM 1993b: 154). The nutrients originate from many different kinds of sources. Some discharges are located on the coastline, while others are situated near rivers with their outflow into the Baltic Sea. Considerable volumes of nutrients also enter the sea from atmospheric deposition. The largest sectors contributing to the eutrophication are agriculture, combustion-driven transportation and waste water from municipalities and industries. These sources contribute approximately one third each.

The concentration of harmful heavy metals in organisms living in the Baltic Sea is considerably higher than in biota from the North Sea. Although the increasing trend seems to have leveled off due to measures taken to reduce the spread of heavy metals into the environment, there are still large volumes contained in sediments. Large influxes of salt water or other disturbances can cause amounts of heavy metals to be released, which cause considerable danger to most living organisms (HELCOM 1993b: 154). The prohibition of PCBs and DDTs in the 1970s resulted in considerably lower concentrations in biota, and observable positive effects in reproduction (SNV 2000a). Some other persistent organic pollutants, however, are still in use, although some of them have recently been regulated in an international convention (Selin 2000).

Almost all investments, changes of practices, technologies, and other initiatives implemented to halt the negative environmental trend in the Baltic Sea have been taken by individual countries adjacent to the Baltic Sea. The creation of the regional Helsinki Commission (HELCOM) on a Finnish initiative and as a result of the Stockholm Conference on Human Environment in 1972 was an important outcome of, rather than the cause behind, the process of increased transnational, environmental cooperation (Hjorth 1992). The HELCOM secretariat has compiled and distributed scientific findings to the contracting parties, and also to some extent influenced national Baltic Sea environmental agendas of the contracting parties. The first Helsinki Convention was formulated in 1974 (ratified 1980) and a revised version was signed in 1992 (HELCOM 1994). Of even more importance was the formulation of the Baltic Sea Joint Comprehensive Environmental Action Programme (JCP) in 1990 (Ringius 1996). The legacy of HELCOM, the Helsinki Convention and the importance of JCP will be discussed at more length in Chapter six.

4. CURRENT RESEARCH

It has been common in the discourse of international relations in the last decade to perceive *regimes* as somewhat of a theoretical umbrella for several kinds of analytical approaches. The structural setting of the interaction is assumed to be determined by the kinds of norms, principles, rules, and procedures relevant in the particular case. Beneath this meta-framework, different theoretical constructions are found, such as for example those emphasizing power or national interests as the main determinants of the outcome of the interaction process.[1] Most scholars in international relations do not, however, use a well specified and distinct theoretical model based exclusively on the assumptions of a particular theoretical tradition, but rather a mix of different ones. This makes it quite complicated to categorize the dominant theoretical constructions of contemporary international relations. The identification of the differences between the main approaches is facilitated if their primary focus of explanation is determined. This categorization is helpful in identifying and highlighting important differences, although most scholars do not use these "ideal" types of theories in their actual analysis. It seems to be possible to distinguish four general theoretical approaches in the contemporary studies of international relations, each emphasizing different primary explanatory factors:

- Interest-based approaches (e.g. rational choice)
- Power-based approaches (e.g. realism)
- Approaches emphasizing cognitive capability (e.g. epistemic communities)
- Structural approaches (e.g. Marxism)

Interest-based explanations place their analytical focus on the interests of the actors. The structural setting is as important as in other approaches, but the point of departure is nevertheless the subjective preferences of the actors. In the actual decision-situation, the outcome is determined by the interaction of the actors as restricted by the structural setting. One of the most prominent approaches in this category is the theory of rational choice.

Power-based explanations similarly emphasize the importance of the actor, but here their capability rather than their interests are the focal point. The outcome is primarily determined by the power-relations between relevant actors. The realism school belongs to this category.

---

[1] A description of this categorization can be found in for example Corrales et al (1999).

Approaches emphasizing cognitive capability stress that it is important for the understanding of international interaction to analyze how different situations might be subjectively perceived by the actors, and that the interaction process itself might change the preferences of the actors involved. The epistemic community approach has been very influential in this category, especially concerning environmental issues.[2]

The final category, structural approaches, is the only one denying the individual actor any essential influence over outcomes. Although individual action is not unimportant, it is seen as the result of the individual's hierarchal position, rather than of his preferences. Theories in this category are consequently deterministic, making understanding of the structural situation sufficient to predict the outcome. Marxists have historically been the most prominent advocates of this approach, but also certain kinds of modern, evolutionary theories belong to this category.

Regarding international, environmental cooperation it is quite clear that three theoretical approaches have been the most extensively used: regime analysis, epistemic communities, and rational choice. The latter two have been located in the categories above, while regime analysis has to be placed in the structural approaches category. The reason for this is that when the regime concept is refined, (i.e. all explanatory factors belonging to theories in other categories are removed), only structural features are left (e.g. norms, rules, conventions etc). It is not the case, however, that regime studies generally are deterministic in their nature, since other theories (i.e. interest-based, power-based, or cognitive theories) are almost always incorporated in order to enhance the explanatory power.[3]

In this study, the rational choice approach has been chosen. The reasons are presented below, after a short review of the other two most relevant approaches: regime analysis and epistemic communities.

4.1. Regime analysis

The Regime analysis discourse commenced with the publication of Power *and interdependence: world politics in transition* by Robert Keohane and Joseph Nye in 1977, and was subsequently refined primarily by Stephen Krasner in *International regimes* (1983), Robert Keohane *After hegemony: cooperation and discord in the world political economy* (1984), and Oran Young in *International cooperation:*

---

[2] Cognitive capabilities have also been emphasized by for example Jönsson (1990) within the framework of negotiation analysis. Negotiations are not, however, central to the understanding of the Swedish environmental assistance, since the decision to grant resources is taken by the Swedish authorities without much formal negotiation. Interestingly, though, the need to focus more on negotiation space and specification of potential efficiency gains has been expressed by for example Underdal (1991). For a recent example of negotiations analyzed using a game theoretic approach, see Fearon (1998).

[3] For a comprehensive treatment of these different approaches, see for example Selin (2000).

building regimes for natural resources and the environment (1989).[4] In addition to these works, a large number of authors applied regime tools in a number of different empirical settings.[5]

Stephen Krasner, in 1983, formulated a definition of an international regime cited in a large number of works:

> [A] set of implicit and explicit principles, norms, rules and decision-making procedures around which actors' expectations converge in a given area of international relations (Krasner 1983:2).

The common way to proceed has been to use this definition, study empirical data from a particular issue-area, and determine whether observed behavior could be described as following particular "principles, norms, rules and procedures".

In the first work by Keohane and Nye, the perception of a regime was very close to that of an institution. What was added was an attempt to put the regime concept on a theoretical foundation in the rational choice tradition, albeit in a rather unformalized manner.[6] The prevalence of regimes could be explained on rational grounds, as a way for the actors to reach common goals while minimizing (transaction) costs and raising costs for free-riding. The persistence of regimes despite changes in relative power was explained primarily by large sunk costs. Regarding the power concept, the most important factor determining the influence of specific actors was not military capability, but rather a combination of power in the specific issue-area under consideration and the importance of the particular issue to the actor.

These rational choice theoretical underpinnings of the regime approach were subsequently lost to a large extent.[7] Later writers have predominantly used Krasner's definition of a regime as a tool to describe different modes of inter-state interactions in different areas.[8] The primary motive behind this study is to explain the reasons behind, and the particular structure of, the Swedish environmental assistance to the Baltic States. This means that the regime approach is not fully adequate. It could

---

[4] In the common property resource (CPR) discourse the empirical focus is set mostly on small local communities. The most distinguished writer in this area is Elinor Ostrom (see for example Ostrom (1990; 1994), Ostrom, Gardner & Walker (1994), and Ostrom & Keohane (1994). The analytical method is, however, very similar to that of regime analysis. It is sometimes stated that regime analysis is undertaken at the macro level, while CPR-analysis deals with issues at the micro level. This is not correct, since they are both at the same analytical level. In both cases, the interaction between "individuals" is studied, although the "individual" in the regime context equals an individual state. The proper micro level deals with the assumptions of individual behavior, while deductions from these assumptions are tested at the macro level, where the outcome of the interaction between the actors is studied. The legitimacy of treating states as individuals will be discussed further later.

[5] A few examples in the area of international environmental cooperation are Young (1991), Bernauer (1995), and Haas (1993). For a review of theoretical aspects on regime theory, see for example Rittberger (1993).

[6] Especially in Keohane (1984).

[7] A few authors have, however, kept the interest base in their explanations, and dropped the regime concept. See for example Vaahtoranta (1994) and Oye (1994).

[8] See for example Young (1982) and Young & Osherenko (1993). For critical reviews, see for example Strange (1982) and Milner (1991).

have been used, if supplemented by the rational choice instruments. Since it is not clear, however, what the regime theory could offer that cannot be interpreted as structural components in a rational choice approach it seems preferable to increase parsimony by using only the rational choice approach.

4.2. Epistemic communities

In the broad category of knowledge-based approaches to international environmental co-operation, the concept of *epistemic communities* has been one of the most prominent. Drawing primarily on traditions in organizational theory, history of science, and sociology of knowledge in general and the continental philosophic tradition of J. Habermas in particular, Peter M. Haas published his analysis of the creation of the Mediterranean Action Plan in *Saving the Mediterranean: The politics of international environmental cooperation* in 1990. Although Haas analyzed a particular issue-area, his theoretical approach has also been applied in other empirical settings.[9] The key issue in this approach is the shortage and unequal distribution of detailed information. An upcoming environmental issue attracts the interest of affected governments and electorates, but it does so in different ways. Although system uncertainty is also an integral part in regime analysis, it is considerably more pronounced in the knowledge-based cognitive tradition. The consequences from for example heavy metal emissions, ozone depletion, or global warming are not easily observable, which forces policy makers to rely rather heavily on statements by scientists. When governments appreciate that there is a problem but they do not know the solution to it, they often tend to turn to scientists for advice (Haas 1990). As the uncertainty makes the subjective interests of the individual states indeterminate, opportunities are given for scientists to influence problem formulation and inter-state cooperation.

The ability to present solution proposals in a unified way is crucially important for the potential influence on governmental decision-making by the scientific group. Such a unified group has been labeled an *epistemic community*. According to Haas, the members of this community share "...an acceptance of a common body of facts, [and] ...a common interpretive framework, or "consensual knowledge", from which they convert such facts, or observations, to policy-relevant conclusions." (Haas 1990:5) If there are large divergences within the group, different decision-makers will most probably be tempted to choose scientific opinions according to their own agendas. It is somewhat unclear whether Haas also includes various kinds of experts, other than scientists as potential members of an epistemic community. He explicitly states that the epistemic community may gain increased influence when some of its members for example become employed in key positions in the bureaucracy (Haas 1990:56). It seems, however, that he means that these former scientists still belong to the same epistemology of science, even after their departure from academia. An individual scientist might become a state bureaucrat, but as long as his/her problem

[9] See for example Corell (1999).

understanding converges with that of the epistemic community, he/she is still a member of it. Otherwise not.

In the case of environmental cooperation in the Baltic Sea region, it is doubtful whether an epistemic community existed in the 1991-1996 period. The potential candidates would have been the scientists connected in various ways with the Helsinki Commission, in combination with segments of bureaucrats in member countries and various NGO experts. Although there was quite a consensual perception of the key environmental problems among the scientists from the West, this understanding was not fully shared by colleagues in the Baltic States and Russia. The lack of an apparent epistemic community was not caused so much by rival scientific opinions, however, as by restricted financial resources. It was simply often not possible for scientists from the CEE countries to attend international conferences, use the latest analytical equipment, and to organize extended research projects. If an epistemic community existed, it was, consequently, far from perfect.

Another reason why the epistemic community approach has not been used in this study is its weak fundamental explanatory potential (Laver 1997). In order to explain outcomes at the macro (international) level, more precise assumptions about the incentives driving the members of the epistemic community would be needed. Even if a clear-cut epistemic community did exist where the members shared a common view of the problem at hand and even of its solution, it is not obvious why these researchers would attempt to alter government policies. They might, for example, see it as more important to maximize their individual reputations among fellow researchers. Most likely, the divergent interests of the other key actors in the issue-area would be of even more importance. Although it is unlikely, a situation where all involved governments, key industrial and other interests share a common view of a transnational environmental problem and its solution is imaginable. This would, however, not guarantee a successful implementation. In most cases the gains from successful abatement projects benefit the actors in varying degrees. It could, for example, be argued that a revitalization of the Baltic Sea would benefit Sweden more than the other coastal states, since more than 60 percent of the Baltic Sea coastline belongs to Sweden.[10] Furthermore, provided that not all actors are pivotal to the realization of the initiative or that there is a disputed scheme to cover its costs, there will be actors with incentives to free-ride, thus jeopardizing a successful implementation.

4.3. The rational choice approach

The rational choice methodology has been one of the most widely applied approaches in political science during the last decades, and is in itself an adaptation of the neoclassical economic theory to the study of politics. As an illustrative example,

---

[10] The share for the other states is: Finland (17%), Estonia (5%), Germany (5%), Russia (4%), Latvia (3%), Poland (3%), and Lithuania (0.6%). Source: personal communication with the Swedish Meteorological and Hydrological Institute (SMHI), 1999-04-21.

the percentage of articles using a rational choice approach in the journal *American Political Science Review* steadily increased from five in 1962 to almost 40 in 1992 (Shapiro 1994). It should be noted, though, that the rational choice approach has been particularly popular in the Anglo-Saxon world. In continental Europe, it has been used with a considerably lower frequency. Rational choice approaches in political science have been used most extensively when analyzing domestic policy. Major research areas have been voting behavior, bureaucratic incentives and performance, and institutional design.[11] It has, however, also been used in international relations by numerous authors, although mainly as a heuristic device rather than a potent analytical tool.[12] It has rarely been used in the issue-area of international environmental cooperation.[13]

The theoretical approach chosen in this book could be seen as a middle way between using rational choice and game theory as heuristic tools and as rigorous theoretical instruments. In order to enhance the accessibility of the text, no formalism will be used. On the other hand, the fundamental premises and requirement of analytical stringency of the rational choice theory will be taken seriously. Without theoretical precision, little of value can be said about potential cause-effect relationships. In other words, this approach is similar to a formal approach in its attempt to be analytically precise. Since this precision is to be achieved without mathematical expressions, however, the verbal argumentation must be careful, and sometimes somewhat lengthy. This careful verbal argumentation ensures internal validity. Needless to say, precision can never be as high in verbal as in mathematical presentations. This loss of theoretical precision is, however, balanced by a higher degree of accessibility. Since the number of readers with deep familiarity of mathematical game theory is quite limited, this urge to increase accessibility is important (Walt 1999).

Scholars representing other approaches, especially in political science and sociology, have repeatedly criticized the formalism and extensive simplifications that characterize most contemporary rational choice approaches.[14] The publication of *The pathology of rational choice: A critique of applications in political science* by D. P. Green and I. Shapiro in 1994 and the response *The rational choice controversy: economic models of politics reconsidered* by J. Friedman (ed.) in 1996, marked the beginning of a very fertile intra-disciplinary discussion over the fruitfulness of the approach.[15] Unlike earlier criticisms, this time it came from two scholars from within the field, and with an undisputed competence in the area of

---

[11] Quite a number of economists have analyzed research problems in the intersection between political economy and political science. A few examples are studies on the role of central banks (Persson 1996) and (Tabellini 1993), Reputation and rational cooperation in finitely repeated Prisoners' dilemma (Kreps 1982), Reputation and imperfect information (Wilson 1982), Evolutionary approaches in game theory (Weibull 1992).

[12] One exception though, is Carraro (1997).

[13] Two recent examples though, are Hanley & Folmer (1998) and Missfeldt (1999).

[14] See for example Hauptmann (1996), Schweers & Levi (1990), Slote (1989), and Etzioni (1988).

[15] For additional comments on the book by Green and Shapiro, see also Cox (1999). Similar issues in the rational choice approach are discussed from a sociological perspective in Coleman & Farraro (1992).

discussion. They were not questioning the approach as such, but rather how it had been interpreted and applied.

One of the most important criticisms by Green and Shapiro was the almost complete lack of, according to their view, thorough empirical testing of the models.

> It [rational choice scholarship] is marred by unscientifically chosen samples, poorly conducted tests, and tendentious interpretations of results. As a consequence, despite its enormous and growing prestige in the discipline, rational choice theory has yet to deliver on its promise to advance the empirical study of politics. (Shapiro 1994:7)

Despite the somewhat indignant response from for example Morris P. Fiorina (1996) and Kenneth Shepsle (1996), Green and Shapiro made an important point. It is quite clear that the bulk of the work within the rational choice scholarship has been undertaken in the areas of model development and theoretical refinement. The achievements in this area are hardly disputed. But there are few examples of "hard testing" of the models, which was acknowledged also by Fiorina and Shepsle. Most models have not been carefully tested against an unbiased data set.[16] This does not mean, however, that they have not been tested against empirical data at all. The issue of empirical testing eventually comes down to an inter-subjective evaluation by the researchers in the field over how convincing the empirical validation of the model is.[17]

The practice among rational choice theorists to illuminate their models by using empirical examples has been misperceived many times as sloppy empirical testing procedures. This is unfortunate. Empirical examples may, or may not, be convincing in regard to model validity. As stated by Jack S. Levy, "..."True or false?" in terms of direct tests of a theory or its hypotheses against the empirical evidence, is not the only question, and social scientists have increasingly shifted to a Lakatosian conception of scientific progress and a demand that theories be tested against alternative theories." (Levy 1997:97) Presumably, a clear-cut distinction cannot be made between adequate empirical testing (quantitative, statistical procedures) and inadequate methods (all other procedures). The enormous influence on the enhanced understanding of political processes despite their lack of hard tests exercised by authors such as Downs (1957), Arrow (1951), Olson (1965), and Hardin (1968) shows this quite clearly. Rigorous statistical testing is certainly one of the most powerful ways to construct a convincing argument, but it is far from the only one (Abelson 1995). Depending on the nature of available empirical data, other "softer" qualitative testing procedures might be preferable in some cases.

In this study, both quantitative and qualitative data will be used. The former will consist primarily of data on the size and number of aid projects, while the latter will be based on a more contextual understanding of the Swedish assistance program. By using different kinds of data in this way, the validity of the conclusions will be higher, and the understanding of the research object more comprehensive.

---

[16] Some exceptions are for example Krehbiel (1991) and McCubbins (1993)

[17] The opinions expressed here to some extent draw upon Fiorina (1996) and Shepsle (1996).

5. THEORETICAL DEPARTURE

One of the most important and critical issues when adopting a game theoretic approach is to carefully define the fundamental unit of analysis. Strictly speaking, the fundamental unit is always the individual in rational choice analysis. All assumptions on the micro level should be made with reference to the individual. There are, however, numerous research areas where the actual fundamental unit is not the individual but some kind of collective. In economics it might be a firm, in election studies a political party or an interest group, and in international relations it is in most cases individual states. The use of states as fundamental analytical units is definitely more problematic than using, for example firms or political parties. It might be argued with reference to firms, political parties and interest groups that these are all exposed to forceful selection mechanisms, making the best adapted most viable. The assumed intentions driving the individuals in a firm, an interest organization, or in a political party are (in this order) less disparate than what can be assumed of a nation state. Additionally, the decision-making structure in a typical firm and in some interest organizations is hierarchical rather than democratic, thus simplifying the transformation from collective to individual preferences. None of these features characterizes democratic states. They are not exposed to forceful selection mechanisms in the same way as the other collectives, they consist of individuals and groups with often very disparate preferences, and they are less hierarchical than firms and many interest groups. This means that it cannot simply be assumed that states can be treated as unitary actors. It has to be shown that *in this particular case*, it is reasonable to imagine Sweden as a unitary actor, because of the high degree of consensus among the key decision-makers. If this had not been the case, rational behavior on behalf of Sweden could not be deduced from the assumption of individual rationality, without severely decreasing the validity of the study.

5.1. The Condorcet paradox

The methodological difficulty with treating collectives as a priori unitary actors has been well known for several decades. As early as 1951, Kenneth Arrow showed that there is no selection procedure that satisfies reasonable requirements in a democratic society, that guarantees that a social preference order can be compiled from a set of individual preferences (Arrow 1951). When trying to arrive at a social order of preferences, the threat of the so-called Condorcet's paradox is always present.[18] The following example illuminates the potential instability in pair-wise voting procedures.

[18] Marquis de Condorcet showed as early as the eighteenth century that the result from a pair-wise voting procedure with more than three alternatives and actors, might be cyclical. This finding later became known as the *Condorcet paradox*.

Assume there are three individuals {A, B, C} who have three alternatives {X, Y, Z} each to choose from. In Table 1 below, the order of preference of these individuals is shown. The best alternative for each individual is at the top.

Table 1. Order of preference of {X, Y, Z} for individuals {A, B, C}.

	Individual		
	A	B	C
Preferences ↑	X	Y	Z
	Y	Z	X
	Z	X	Y

Let's say we match X versus Y in the first round, then X will beat Y since A and C prefer X to Y. The winner X is then matched against Z, and Z will win since B and C prefer Z to X. *The winner is Z.*

Then we start with Y against Z, which will give Y as the winner, since A and B prefer Y to Z. In the following round, Y will lose to X, which means that *X will be the winner.*

Let us finally put X against Z in the first round. Then Z will win, but lose to Y in the final round, which means that *Y is the winner.*

This example shows that the winning alternative might be arbitrary. It all depends on the order the alternatives are matched against each other. The power of the agenda-setter might thus be considerable. It is also clear, however, that the order of preference by no means *has* to be cyclical. If the individual ranking orders were identical, for example, the social preference order would be identical to the individuals', and the problem would evaporate.[19] There are of course several other procedures besides pair-wise voting to aggregate a social preference order from individual preferences. As shown by Kenneth Arrow in the so-called *impossibility theorem*, however, they all exhibit some problem not satisfactory in modern democracies.

Because of the crucial importance of showing that only small differences exist between individual preferences within a group and thereby reduce the relevance of Condorcet paradoxes, a considerable effort has been made in this study to show that the views on the Swedish environmental assistance to the Baltic States have been very homogeneous. In Chapter four, it will be shown that there has been an extraor-

[19] Although the voting cycles have been studied very extensively in theory, systematic application on empirical data rather than using casual examples has been quite rare. One recent example, though, is Lagerspetz (1997), where cyclical preferences in several presidential elections in Finland are shown to have had a significant influence over the eventual outcome.

dinary convergence among the preferences of the Swedish political parties. Therefore, the treatment of Sweden as a unitary actor is reasonable.[20]

5.2. Individual and social preferences

It has traditionally been common in international studies to ignore the inherent difficulties involved when treating states as unitary actors. The focus has been set on systemic factors, and the interacting states have often been analytically perceived as "black boxes". In the last decade, however, much interest has been focused on the search for domestic explanations to observations at the systemic level.[21] In a game theoretical approach, the focus on both the domestic and the international scene is rather natural.[22] This was particularly highlighted by Bueno de Mesquita and Lalman in their influential book *War and reason: Domestic and international imperatives* (1992). They defined an *international interaction game* where the explanatory focus was put on domestic restrictions and incentives as well as on systemic structure: "In the constrained, domestic version of the international interaction game, domestic imperatives, rather than international considerations alone, play a prominent role in foreign policy choices" (Buena de Mesquita & Lalman 1992:266).

In this study it is assumed that the utility of the actors (states) is only affected by domestic gains and losses. The structure of the strategic situation is then formulated as a game at the macro level. As the model is validated against behavior at the macro level – the outcome of the interaction process between the involved states – it will be possible to propose what Michael Laver has labeled a *potential fundamental explanation* (Laver 1997). By separating the micro and macro level and keeping all assumptions at the micro level, hypotheses can be *deduced* at the macro level, which can then be validated against empirical observations. This procedure reduces the risk for circular arguments. Although Michael Laver presents this procedure as something almost new, it is the traditional, positivistic methodology still used by virtually all researchers in natural science as well as in economics. The powerful postulate, making it possible to formulate hypotheses about cause-effect relationships, is the assumption that the actors will maximize their utility. This assumption is made in all rational choice approaches, as well as in neo-classical economics. The fundamental meaning of this postulate is that the individual is supposed to act instrumentally rationally to reach his/her goals, whatever these might be. This is what the rationality assumption in rational choice is.

In the rational choice discourse, a distinction is sometimes made between *thin* and *thick* rationality (Tsebelis 1996). The thin rationality assumption is defined as

[20] If this had not been the case, it would have been necessary to formulate the model as a two-level game, with a domestic game defining the preferences of the actors in the international game. For an example of application of two-level games, see Tsebelis (1996).
[21] See for example Putnam (1988), Buena de Mesquita & Lalman (1992), Downs et al (1995), and Gourevitch (1996).

only instrumental rationality. The thick variant, in addition to instrumental rationality, consists of a postulate about egoistic behavior (i.e. the utility to other actors is not part of the utility function). Karl-Dieter Opp (1999) has suggested another distinction, where *narrow* and *wide* conceptions of the rationality assumption are sub-groups of the *thick* version.

The definition of rationality is kept as simple as possible in this study. Assumptions on self-interest or altruism are therefore excluded from the definition of rationality. When the concept of rationality is mentioned henceforth, if nothing else is specifically stated, it thus refers to thin rationality.

The rationality assumption has been perceived as quite unrealistic to many political science scholars outside the rational choice discourse. The empirical validity of the utility maximization assumption has been challenged by, for example, proponents of the so-called *prospect theory.* Prospect theory could be seen as a special branch of the rational choice approach, and has primarily been used by psychologists.[23] It has been noticed by a number of scholars that generally people do not seem to maximize expected utility when the probability of an event is very small or very large.[24] This empirical observation has been theoretically elaborated in prospect theory, where the assumption of expected utility maximization has been abandoned. It is not yet settled whether this empirical observation can be explained using the expected utility maximization assumption or if a modified variant such as for example that of prospect theory is necessary. As such a modification would in all probability make the approach less parsimonious, it would be preferable to use the traditional approach, provided the empirical anomaly can be solved. In this study, however, the use of maximization of expected utility is not problematic, since the focus has been set on behavior at a more abstract level.

Regarding altruistic behavior, it should firstly be noted that there is no inherent contradiction between assuming altruistic behavior and the rational choice approach. Models can be formulated where the actors are assumed to have altruistic preferences. One has to define, however, precise statements about the assumed preferences. Ad-hoc saving of the model by also referring to preferences not assumed from the beginning deprives the analysis of its explanatory power. It has been suggested by the sociologist Lars Udehn that it might be assumed that the individual has not one but several utility functions (Udehn 1996). By assuming this, one would move one step down, and explore whether it is possible to collapse these functions into one (in analogy with the discussion of the formulation of social preference orders above). This approach seems possible to pursue, although it would complicate the analysis considerably. It would probably be of more interest in a situation where

---

[23] For an excellent comparison between rational choice theory and prospect theory, see Kahneman (1986). Levy discusses the differences between rational choice and prospect theory in the context of international relations (Levy 1997). For empirical applications of prospect theory, see for example Weyland (1996) and Berejikian (1995).

[24] One famous example of this is the *Allais paradox*, where experiments seem to consistently show that people do not always maximize expected utility. The Allais paradox is described for example in Yamaguchi (2000).

there are only a few individuals involved, and not actors consisting of collectives. Therefore, and because it has not been explored adequately, this approach will not be referred to again in this study.

What is not compatible with the rational choice approach is the absence of instrumental rationality, irrespective if the individual preferences are assumed to be self-interested or altruistic. The critique against the assumption of instrumental rationality has to a considerable extent been caused by a misunderstanding. The relevant null hypothesis to evaluate this assumption against is *not* empirical observation, but rather alternative assumptions on individual behavior. The issue is not whether the individual can be assumed always to act instrumentally rationally, but rather whether most individuals will do so more often than they follow any other systematic rule of behavior. If instrumental rationality is not assumed, one would have to choose to postulate either that most individuals act randomly without any relation between ends and means, or that the guiding principle is to act systematically to avoid a correlation between ends and means (Tsebelis 1996). Neither of these assumptions seems more reasonable than the postulate of instrumental rationality.

5.3. Gains and costs from joint environmental projects

The Swedish assistance program vis-à-vis the Baltic States consists not only of environmental support, but also of initiatives targeted to support, for example, general institutional capacity, democratization, economic development, and security. It is important to note that the method used in this study does not require any assumptions on how the assistance has been allocated *between* environmental assistance and other parts of the support program. This means that it is reasonable to assume that the allocation within the area of environmental support will be primarily guided by environmental considerations, although the over-all allocation between different sectors might be guided by other concerns.[25] More precisely, it is hypothesized that Sweden prefers projects with a high collective good content, that is, projects benefiting not only the recipient country but also Sweden. Projects with a high private good content, that is which primarily benefit the recipient country, will consequently be less preferable to Swedish authorities.

The standard assumption in neo-classical economics that individual *tastes* can be treated as constant during the time period under study has sometimes been criticized as unreasonable. However, as shown by, for example, the influential economists G. J. Stiegler and G. S. Becker, this assumption can in most cases be defended (Becker 1977). In fact, they suggest that tastes can, and should, be treated as stable over time and similar among all individuals. What might at first glance be seen as changes in tastes can, after a closer look, be found to be changes in information levels and

[25] The choice of environment projects might also to some extent be influenced by, for example, attempts to support domestic industries in the donor country. This will be discussed somewhat further in Chapter seven.

prices. "...they [assumptions of differences in tastes], along with assumptions of unstable tastes, have been a convenient crutch to lean on when the analysis has become bogged down. They give the appearance of considered judgment, yet really have only been *ad hoc* arguments that disguise analytical failures." (Becker 1977:89)

In this study, it is not necessary to go as far as Stiegler and Becker regarding the general applicability of stable and homogenous preferences. As will be shown in Chapters four and five, however, the assumption of stable, national environmental preferences in the period 1991-1996 is reasonable and in accordance with empirical observations.

Another standard assumption in rational choice is that actors are assumed to be forward-looking. Prior events are only relevant if they affect future outcomes. By choosing between available strategies in a specific situation, the actors maximize their expected outcomes according to their subjective preferences. It is not given, however, that the actors have to be forward-looking. An example of backward-looking actors is the so-called stochastic learning model, where the actors are assumed to adapt to their environment by changing strategies according to prior events. By definition, such an actor cannot perceive casual relationships since the learning process is inductive. Instead, the actors observe only correlations and adapt their behavior according to these observations. For examples on how the stochastic learning model has been used, see for example Macy (1989) (on a potential solution to the Prisoner's dilemma), Kanazawa (1998) (on a potential solution to the paradox of voting), and Gautschi (2000) (on the potential importance of the "shadow of the history"). In this study, however, the most common approach is used, where the actors are assumed to be forward-looking. The most important reason for this choice is the fact that the Swedish CEE assistance was to a certain extent something new. It therefore seems less reasonable to explain behavior by past experiences.

It is assumed in this study that the net gains to the actors from a particular joint project are determined by four parameters: environmental benefits, project cost, size of the foreign grant and transaction costs. These costs and benefits are distributed in a specific way in each particular project. Although the use of only four variables is a gross simplification, this assumption makes the analysis of the incentives facing the actors much more tractable. It should be noted, that the values of all variables are *expected* values. The only exception is the foreign grant, which is known by the donor, but not by the recipient country.[26] This is the only value that can be established by the donor beforehand.[27]

The environmental benefits from such a project can analytically be seen as con-

---

[26] As will be discussed further in the analysis in the last chapter, a praxis might, however, have developed where the recipient country has become well informed about the expected size of the grant. Even in this case though, the donor country might at any time choose to change its proportion of financing.

[27] The donor country could choose to alter the proportion of financing during the implementation of the project. This, however, does not alter the choice situation faced *before* deciding whether to support a particular project, or not.

sisting of two different parts: one that is private, and thus benefiting only the country where the project is undertaken (i.e. one of the Baltic States) and one that is collective, thus benefiting both countries equally. In other words, if the issue concerns a joint project implemented in one of the Baltic States, the benefit accruing to this Baltic State equals the sum of private and collective gains, while Sweden can only gain from the collective part. It is assumed that the collective part benefits both the recipient and donor countries equally. It should be noted, that this does not mean that the benefits emanating from a particular environmental project are either private or collective in the typical case. Rather, such benefits consist of a mix of private and collective gains. Sometimes environmental benefits are primarily collective to their nature, and sometimes primarily private. The analytical advantage achieved by separating private and collective benefits derived from the project will be discussed in more depth in Chapter two.[28]

The term project cost is defined as the full cost related to the implementation of the project. The distribution of this cost among the different actors varies depending on the type of project. It is assumed for analytical clarity, that the recipient country covers the full project cost, but that foreign grants reimburse some of these costs. In all joint projects analyzed here, Sweden covers a specific part of the particular project. In the typical case, this parts amounts to approximately one quarter of the total costs. Another part is always covered by the Baltic State itself. Sweden never carries the full cost of the project. The reason for this will be discussed in Chapter two. In many cases, additional donors and/or international financial institutions cover significant parts of project costs.

Transaction costs are assumed to detract utility from joint projects, since it can be assumed that project administration becomes more complicated, as the number of cooperating partners increases. In other words, the transaction costs referred to here are costs that are directly related to the fact that a foreign aid donor becomes involved. Other kinds of transaction costs not related to the introduction of a provider of foreign aid are not included, since these transaction costs are identical irrespective of whether it is a domestically or jointly financed project. This means that from the grant originally provided by the donor, the recipient country receives a somewhat smaller amount. This amount is labeled net grant. A particular project thus becomes somewhat more expensive when it is jointly financed compared with a situation where only domestic resources are used. Despite the different context, the set of variables presented here bears some resemblance to the so-called *gravity model*, which is used to predict transnational trade flows. In the gravity model, geographical distance is often used as a proxy for transportation and transaction costs (Mansfield & Bronson 1997). Here, however, transportation costs are included among other costs in the transaction costs term. As the geographical distances from Sweden to all three Baltic countries are similar, no distinction is made between them on these grounds. In other words, although geographical distance is presumably positively

---

[28] A similar Olsonian separation of private and collective utility has been made in the context of analyzing behavior of American, environmental citizen groups, see Lowry (1997).

correlated with transaction costs, all three countries are affected in similar ways. The relevant distinction in this context is thus between projects implemented in Sweden (no transaction costs, according to this model) and those put into practice in the Baltic States (positive transaction costs), and not whether it is Estonia, Latvia, or Lithuania that is the co-operating partner.

In Table 2 below, the assumed distribution of costs and benefits between donor and recipient country is summarized.

Table 2. Distribution of joint project benefits and costs

	Benefits	*Costs*
Donor country	Collective environmental gains	Grant (incl. transaction costs)
Recipient country	Collective environmental gains Private environmental gains	Project costs – net grants

It should be noted that net benefits from a domestic environmental project implemented without foreign assistance in either country could be analyzed using the same scheme as in Table 2. The foreign grant variable would then be zero, and according to the definitions above, so would transaction costs. This is an important part of the model laid out here, since the relevant alternative to a joint project for each actor is to choose a project financed by purely domestic resources. This is the relevant opportunity cost for choosing to implement a joint project. In other words, for a joint project to be implemented, net benefits to each actor must be larger that what could have been achieved by implementing domestic projects, financed by the resources that otherwise would have been invested in the joint project. The financial resources available to the recipient country for such an alternative project would be the resources otherwise spent on the joint project plus the grant, excluding transaction costs. For the donor country the available means would be equal to the size of the grant. Total net gains to each of the countries from a domestically financed project would be private and collective benefits from these respective projects, plus collective benefits emanating from the project implemented in *the other* country. Since both countries gain equally as much from collective benefits, as defined above, these gains are common to both countries from both projects.

In the next chapter, the theoretical framework introduced here is further elaborated. The central task is to illuminate how the most important driving forces behind the Swedish environmental assistance to the Baltic States can be analytically treated. In the following chapters a detailed description is presented of the background and evolution of Swedish foreign assistance and international environmental cooperation (Chapter three), a portrayal of the high degree of consensus in Sweden regarding the environmental assistance to the Baltic States (Chapter four), an account of environmental preferences in the Baltic States in the first half of the 1990s (Chapter five), and a description of the content and administration of the Swedish support (Chapter six). By giving this comparably detailed empirical orientation, it will then be possible to present conclusions on the major tenets of the Swedish environmental assistance to the Baltic States.

The methodology adopted, where quantitative data are used, but further interpreted in the light of more qualitative empirical data is similar to the approach suggested recently by Karl-Dieter Opp (1999). This should be contrasted to, for example, econometric models where mathematical modeling is applied more extensively, and more precise empirical data are used when such are available (e.g. data on market prices).[29] Since the variables used here cannot be measured directly, the theoretical framework has to be able to handle available data at a more abstract level. Shortage of relevant and precise data is, in fact, a problem facing most quantitative studies of transnational, environmental issues (Missfeldt 1999). The so-called *modeling by example* approach has therefore been used in order to capture key variables, describe their possible relationships with each other, and generate interesting hypotheses, interpreted in a quite unformalistic manner.[30] The ambition is to validate directions and tendencies predicted by theory. It will, for example, be established whether the Swedish selection of joint projects is compatible with the tendencies and directions predicted by the model.

---

[29] A recent example of this more formalistic approach applied to environmental issues can be found in Hanley and Folmer (1998).
[30] The modeling by example approach is described in, for example, Rasmusen (1997).

CHAPTER 2

MODELING SWEDISH ENVIRONMENTAL ASSISTANCE

1. CHAPTER OUTLINE

In this chapter, a theoretical framework for the analysis of Swedish environmental support to the Baltic States is presented. First, it will be shown that transnational cooperation has been perceived as necessary by the riparian states' governments in order to rehabilitate the environmental status of the Baltic Sea. This perception seems to imply that collective goods exist. Otherwise, private incentives, in combination with established property rights would, strictly speaking, be sufficient, and other institutional mechanisms would be superfluous. The reasonability of the assumed existence of collective goods will be discussed and analyzed.

Some definitions are needed: The term *private* refers to the actor. Private benefits hence are attributable to a particular actor and not to a collective. A *good* is any object, service or other tangible or intangible item that brings a positive utility to the relevant actor. A *bad* is in a similar way anything that detracts utility. The emphasis is on propositions on the margin, rather than on finding point predictions. The actor in this study is the nation-state, which means that private benefits are accrued to a particular nation-state.

Having analyzed to what extent different environmental aspects of common goods/bads are present, the theoretical framework is used for the derivation of plausible hypotheses, where the primary aim is to analyze whether Swedish self-interest can be perceived as an explanation to observed behavior. An important aim in this chapter is moreover to generate complementary hypotheses for further discussion and analysis. In Chapter seven, the hypotheses derived here are evaluated against a compilation of empirical data on environmental assistance projects implemented in Estonia, Latvia, and Lithuania. The contextual empirical data presented in Chapters 3-6 and to a lesser extent comparative data from other major providers of environmental assistance will be used to further discuss Swedish environmental support to the Baltic States.[31]

---

[31] The two most important bilateral donors of environmental aid to the Baltic States apart from Sweden have been Denmark and Finland. A review of the Danish aid policy can be found in Ringius et al (1996). The Finnish experience is described and analyzed in Henttonen (1996) and Hiltunen (1996).

2. PERCEPTION OF COLLECTIVE GOOD CONTENT

The need for extensive cooperation between the countries adjacent to the Baltic Sea has been taken for granted by the Swedish government, as well as by all the political parties. The primary objective in the aftermath of the dramatic political changes in Central and Eastern Europe (CEE) in the late 1980s was to improve Swedish national security. In the Government bill of 1994/95, where the strategy for the cooperation with the CEE countries for the following three years was outlined, it was stated that:

> It is of the highest priority for Sweden to primarily develop the common security in its immediate surroundings, around the Baltic Sea, and in the Barents Sea (Government bill 1994/95:160 p. 3).

The emphasis on common security is in line with earlier Swedish security policy, not only in relation to neighboring countries, but also as a general and ideological commitment. National security should be sought not only by increasing domestic capability but also at the systemic level, in terms of direct cooperation as well as adherence to international rules of conduct. Sweden has, for example, traditionally been a strong advocate of a strengthened UN mandate. This standpoint certainly reflects the fact that a small country like Sweden is more vulnerable to weak international codes of conduct, than a larger country that can rely on its own capability to a larger extent. But it also reflects an appreciation of the possibility of reaching joint security gains. International regulation of security issues may be beneficial to all, although there might exist temptations for an individual state to defect from existing agreements or to refuse to enter new ones.

A regional environmental organization (HELCOM) was created, based on a convention drafted in 1974. This originally Finnish initiative was supported by Sweden, as well as by the other western states around the Baltic Sea, as cooperation was perceived imperative to a successful rehabilitation of the Baltic Sea. The first convention was put into force in 1980, and in 1992, all contracting parties signed the revised version.

The great political changes taking place by the end of the 1980s in the CEE countries opened up new opportunities for environmental cooperation. Without depreciating the work of HELCOM, more attention was now given to actual implementation of environmental undertakings in the Baltic States. National measures to improve the environmental quality were a purely domestic issue under the auspices of HELCOM. Therefore, increased activity was most easily attained on a bilateral basis, where other actors could be involved as co-financiers. These new kinds of cooperative activities now became considerably more comprehensive.

3. DEFINITION OF A COLLECTIVE GOOD

In this section, the analytical separation of collective goods into three different dimensions - geographical scope, degree of rivalry in consumption, and excludability -

is described. It is then shown that for the purposes of this study, the most productive approach is to concentrate on the geographical scope dimension.

According to the approach used by Nash (1981), all goods can be located in a three-dimensional space, depending on their *geographical scope, degree of rivalry in consumption* and *excludability*.

The geographical scope variable measures whether a particular good is common to countries in a region, on a continent, globally, or to any other set of actors. The climate convention, for example, is an attempt to combat the decay of an approximately global collective good, a healthy atmosphere. The protection of the Rhine River from excessive pollution of chloride on the other hand, is primarily a concern for the riparian countries Germany, France, the Netherlands, and Switzerland (Bernauer 1996).

The degree of rivalry in consumption captures to what extent a particular good is still possible to utilize even after someone else has already used it. The value of a gain in knowledge on the toxicity of chlorinated carbons, for example, is generally not affected if additional actors share these new findings. This type of knowledge thus exhibits a low degree of rivalry in consumption. For example, utilization of mineral resources from the seabed on the other hand, directly affects potential exploitation by other actors and is therefore characterized by a high degree of rivalry in consumption. In many cases, an item does not belong to any of these extremes, but is located somewhere in between on a continuous scale.

The excludability variable, finally, refers to whether it is possible to exclude certain individuals from consuming a good, or to be subjected to a bad. The distribution of clean drinking water, for example, can rather easily be reserved for suitable sets of consumers. It thus exhibits a high degree of excludability. The protection of the ozone layer, on the other hand, will serve most people on an even basis, at least in a particular region. Specific subgroups of individuals are not easily excluded from this service, or particularly exposed to the lack of it. Therefore, the benefits emanating from the ozone layer exhibit a low degree of excludability. Strictly speaking, it is more a question of the cost of exclusion, rather than if it is possible or not. Concerning constraints on instrumental goal seeking, Robert Nozick has stated a direct analogy to this standpoint: "Constraints may be a matter of degree, more or less difficult or costly to violate. We might better construe a constraint as a *gradient*. (How difficult is it to move in that direction, how much energy does it take to move or stay there, how costly is it in terms of time or resources, how much resistance must be overcome?)" (Nozick 1995:165). The common vocabulary will be used here, however, for the sake of simplicity and transparency.

In this study, the most critical dimension concerns geographical scope. Although most categories of environmental goods are characterized by rivalry in consumption to some extent, this is not of relevance if, for example, Sweden is not affected by a particular environmental initiative implemented in one of the Baltic countries. Irrespective of where the effects from this particular initiative are located on the rivalry dimension, Sweden will only be affected if these effects have sufficient geographical scope. If they do have sufficient scope, the incentives facing Sweden will not be affected so much by rivalry issues as by the actual effect in itself.

The excludability dimension is for similar reasons not relevant here. The notion of excludability has most commonly been used to analyze the potential for reconvening investment costs. If an investment is undertaken where fees from users are supposed to recover investments costs, it is of pivotal importance that those not paying can be excluded from the provided services. Although Sweden can be excluded from benefiting from, for example, investments in fresh-water facilities in the Baltic countries, this is not really the issue when referring to the excludability potential. Rather, the important thing is whether not-paying users in the Baltic countries can be effectively excluded. To Sweden, the important issue is once again that of geographical scope, namely to what extent Sweden is affected by environmental undertakings in the Baltic States. The focus of our interest will therefore henceforth be on the issue of geographical scope.

It will be assumed that the potential utility derived from all goods analytically consists of two fractions, one private and one collective. Each of these fractions may be between zero and one, and they always add up to one. If the private part is zero, the good is perfectly collective, and if the collective part is zero, the good is perfectly private. It does not matter to a particular actor whether the utility is derived from private or collective sources, as each actor only cares about the sum of the private and collective utility accrued to him. A project implemented in one country can only benefit other countries when the collective good part is larger than zero. All private parts by definition benefit only the country where the project is implemented.

4. THE COST OF ENVIRONMENTAL INVESTMENTS

The seminal works of Mancur Olson *The logic of collective action* (1965) and Garrett Hardin *The tragedy of the commons* (1968) incited a huge interest in the analysis of collective goods. These studies were almost exclusively attempts to explain the lack of cooperative behavior or, alternatively, to suggest arguments to account for observations of cooperation.[32] The validity of these models relies heavily on prior assumptions of the existence of collective goods. It is an apparent danger that rational behavior is explained by some kind of altruistic preferences, while it may very well be the case that the "collective good" in question has significant aspects of private rather than collective nature. If this is the case, the inference of altruistic preferences from the observation of cooperative behavior may not be valid.

The opportunity cost of additional environmental investments is of pivotal importance when comparing national, environmental strategies between different countries. It is in most cases inadequate to draw direct inferences concerning national environmental preferences from a comparison between relative spending on environmental issues. It is generally assumed that the marginal utility from income

---

[32] Some of the often-cited authors in the former category include Taylor (1976), Weingast (1984), and Snidal (1991), while Ostrom (1994) and Axelrod (1984) are examples in the latter category.

and most individual commodities is decreasing (Hirshleifer & Glazer 1992). Depending on the level of income, the demand for specific goods might vary considerably. It is for example often assumed that commodities for basic needs such as food, shelter, and security have low income elasticity, while less essential commodities have higher elasticity. Benefits from most kinds of environmental projects can be assumed to fall in the latter category. At the aggregate level, this means that the opportunity cost of environmental projects is negatively correlated with GNP.[33]

The intuition behind the decreasing opportunity cost of environmental projects is simple. The basic needs necessary for survival have to be satisfied first of all. When, or if, the income rises, less essential needs may be addressed. The point to be made here is that if a lower percentage of GNP is allocated to environmental issues in the Baltic States than in Sweden, this does not necessarily mean that environmental concern is intrinsically lower in the former than in the latter. It might be, but to be able to compare Sweden with the Baltic States on this issue, information on the relevant opportunity costs in Sweden *at the same GNP level as in the Baltic States* would be needed. As these kinds of counterfactual data are not available, one can only speculate about which country is the most environmentally concerned, in terms of resource allocation to environmental issues. A historical comparison between the Baltic States today and Sweden some decades ago, when the Swedish GNP was more similar to that of contemporary Baltic States, would not be valid, since a number of external factors relevant to the national agenda setting certainly vary over time.

It should be noted that the intention here is *not* to compare the costs arising from different projects in Sweden and in the three Baltic States respectively. When it is stated that the marginal opportunity cost can be assumed to be higher in a Baltic country than in Sweden, it is because the GNP is lower in the former than in the latter. In other words, the comparison is with a contra factual situation within the same country, rather than between different countries. The existence of diverging opportunity costs is consequently purely theoretically derived.

5. COLLECTIVE ACTION ON COLLECTIVE GOODS

In this section, the unanimous support requirement in the HELCOM recommendations process is shown to have decreased the incentives for free-riding. At the same time, however, it has also given the least determined actor the power of veto over the level of collective ambition. Prior to the institutionalization of Swedish CEE

[33] This assumed correlation should not be confused with the so-called *Kuznets Curve*. The Kuznets Curve depicts a relation between GNP and the state of the environment. It could very well occur that despite increasing GNP, the state of the environment would deteriorate even with increasing levels of investment in the environmental sector (i.e. if the declining opportunity cost for environmental investments does not compensate for increased pollution caused by higher economic activity). What is stated here, however, is only that there is a negative correlation between opportunity cost for environmental investments and GNP. Thus it does not state anything about the state of the environment. For a recent work on the Kuznets Curve, see for example Vogel (1999).

assistance in 1989, environmental protection and abatement with relevance to the Baltic Sea was almost exclusively financed by national means. In the more affluent Nordic countries, large amounts were invested especially in the prevention of point source pollution (e.g. the construction of waste water treatment plants and adequate sewage networks). The HELCOM process was the main joint initiative in the region. Although the focus in this study is on aid projects executed in the 1990s, two features of the earlier HELCOM era are worth mentioning.

According to the Helsinki Convention unanimous support is required to make a recommendation valid. Rules that did not grant national veto would not have been acceptable to the contracting parties (Hjorth 1992). Although the distribution of costs and benefits may vary considerably, all kinds of international conventions can be interpreted as collective goods benefiting all signatories (albeit often to various extents). Since no actors are forced to sign the convention, the fact that the document is signed implies that all parties expect to gain from accepting the agreement. This does not preclude the possibility that a particular party is in some way induced to sign the convention, for example by threats of retaliatory action. As the decision to accept or reject the convention is based on an over all assessment of the utility, acceptance stills implies that the actor expects a net gain from signing the convention. The signing of an agreement, however, does not per se imply that action will be taken in accordance with the convention. The only statement made here, is that an expectation of a positive utility gain from the actual signing of the document exists.

The common wisdom from collective action analysis is that all jointly financed collective goods will make the actors tempted to free-ride. Even if a single actor stands to achieve a positive utility from the undertaking, he might choose not to take part in it, counting on the assumption that other cooperating parties will. Whether the collective good will be produced or not, depends on whether the fraction of actors with a positive utility is sufficiently large to bring about the implementation, *and* whether a self-enforcing mechanism to detect and punish free-riders has been institutionalized (Ordeshook 1986). If, however, these requirements are not met, the collective good may still be attained if every actor's participation is necessary for its realization. That is, if all actors are *pivotal* (Shepsle 1997). Apart from limiting free-riding in this way, an institution where all actors are pivotal to any particular agreement may put some pressure on the so-called laggard states to also adopt recommendations that entail some net utility loss from the recommendation itself.[34] The institutional design of the Helsinki Convention relies on the mechanism of making

[34] Observations of similar kinds have been made by several authors in the contexts of economics: “”The first-best can be achieved because … discontinuity … makes every player the marginal, decisive player: if he shirks a little, output falls drastically and with certainty.” (Rasmusen 1997:215) A number of applications in different contexts but all facing similar problems are cited in Rasmusen (1997: Chapter 8).

all actors pivotal to the acceptance of every specific recommendation.[35] The requirement of unanimous support for all recommendations is not the only conceivable option. From a strict individual point of view, it might certainly be beneficial under certain circumstances to adopt recommendations also when only some of the contracting parties have come to an agreement, thereby letting one or a couple of actors free-ride, or for example adopting dual standards for different groups. In this way, those most interested in a particular initiative may still not be forced to accept inaction. This could be a way to safeguard against the demise of, at least in the short run, Pareto-sanctioned proposals.

There is, however, no guarantee that such an institution will not unravel due to free-riding. When there is at least one actor who is not pivotal and the identity of this actor is not given, the other actors have incentives to pick the role of the free-rider. Since it can be assumed that each actor reasons along the same lines, the outcome will ultimately be that no one wants to sign the particular recommendation. One way to interpret the choice of unanimous consent as a prerequisite for the adoption of a recommendation in the HELCOM process, is thus as a mechanism to impede free-riding. The weakness of this construction, however, is that the scope of the interaction between the contracting parties is determined by the preferences of the least ambitious one.

It should be noted that side payments are assumed not to be allowed. Under most conventions, such mechanisms are not used, even if exceptions can be found (e.g. the Convention on the ozone layer). Side payments were not used under the HELCOM regime before the dissolution of the Soviet Union. The primary tenet of the cooperation thus was the pivotal role of the actor with the lowest denominator standpoint. In the words of P. T. Hopmann, the "complexity of multilateral negotiations has plagued negotiators who have frequently found [these] negotiations highly frustrating and too often tending toward lowest-common-denominator agreements with little substantive content." (Hopmann 1996:270)

6. POTENTIAL GAINS FROM BILATERAL SUPPORT

In this section it is shown that there may exist Pareto-sanctioned investments in a Baltic State, where Sweden covers part of the cost.

At the time when the Moscow-based political control over Poland ceased, and the former Soviet Union states of Estonia, Latvia, and Lithuania became indepen-

---

[35] There are several other conceivable mechanisms designed to curb free riding that have been analyzed in relation to international, environmental cooperation. By attaching issues in other areas, cooperation can be enhanced by so-called *issue linkage* (Whalley 1991), in *matching* approaches one country takes the first step and thereby induces others to behave more cooperatively (Barrett 1995), *social norms* have by some authors been found to enhance cooperation (Hoel & Schneider 1997), *tit-for-tat* and *trigger* strategies are game theoretic strategies that generally increase cooperation (Cesar 1994 and Barrett 1995, respectively). Most of these mechanisms, however, are only applicable if an iterated games approach is used.

dent, new structures for interaction between the countries along the Baltic Sea rim evolved. The opportunities for engagement in more far-reaching joint action initiatives were drastically augmented. These opportunities were on the systemic (regional) level and taken advantage of primarily through the formulation of the Joint Baltic Sea Action Program (JCP). The political restrictions were now considerably relaxed, but the structure enabling the least ambitious actor to decide the extent of the activities remained unchanged. One way to circumvent this constraint was to engage in bilateral cooperation, without sacrificing the achievements made at the systemic level.

The key to the understanding of the gains from bilateral agreements lies in the recognition of mutually beneficial exchange. In microeconomics this potential is conceptualized in the market mechanism, while the same phenomenon is labeled *value creation* in negotiation theories (Kremenyuk 1991). The term value creation may be somewhat misleading, since it is hard to see how value actually is *created.* Rather, the potential value from transactions is always there and the issue is whether the actors are able to actually realize these values or not. In the international negotiations vocabulary, this concept may be interpreted as the point when the negotiators are reaching an agreement, and thereby ”create” value to their principals. In the political sphere, this notion is captured by, for example the, concept of logrolling. The idea is quite simple; when two actors evaluate the utility derived from certain goods differently, an exchange of goods between them increases the utility to both. While the simple form of this utility-enhancing process amounts to some kind of barter, the modern variant utilizes the exchange of money. The use of monetary flows instead of barter trade offers a dramatically increased opportunity to partition resources, and a common means of exchange.

In the realm of transnational environmental cooperation, diverging valuations (i.e. opportunity costs) of environmental investments similarly open up a potential for mutually beneficial exchange. Swedish financial resources can be exchanged for environmental investments in the Baltic States, which potentially can make both parties better off. It should be recapitulated that these differences do not stem from any differentiation in intrinsic preferences concerning environmental issues. Such differences might exist, but this issue is not examined in this book. Here, the focus is on explaining the existence and character of the *Swedish* environmental support to the Baltic countries.[36]

In order to understand the rationale behind environmental assistance to a poorer county, consider the following: When a particular environmental investment is implemented, resources have to be used to finance this investment. These resources could have been used elsewhere, which means that the relevant measurement of the cost of this environmental investment is its opportunity cost. The size of this opportunity cost is obviously linked to the size of the economy. The more abundant

---

[36] Environmental attitudes in the Baltic region have been studied by, for example, Gooch (1995) and Szrubka (forthcoming).

national resources are, the smaller is the sacrifice when implementing the environmental investment, in terms of other opportunities forgone. In a country less economically developed, many needs are still not met, which means that the sacrifice will be considerably larger. It follows, then, that the richer the country is, the more resources have been spent on environmental investments, even if the importance of environmental issues is identical in both countries. It can furthermore be reasonably assumed that the marginal environmental return from additional investments is decreasing. The most urgent needs where the environmental gains are the greatest are addressed first, which means that additional investments will be comparably less beneficial. Assume now – as an ideal case – the wealthier country (e.g. Sweden) considers making an environmental investment, where the benefits in terms of environmental enhancement are purely collective. It then seems clear, that there is a rationale for Sweden to invest in the poorer country (the Baltic State) rather than in Sweden. The environmental gain is identical irrespective of the country in which the investment is undertaken, since the benefits are purely collective. The cost of the investment might, however, be higher when it is implemented in the Baltic State. As was noted above, transaction costs, as defined above, make total costs higher in joint projects than in projects financed purely with domestic resources. The conclusion is then that provided that the difference in environmental gain between investing in the Baltic State rather than in Sweden is larger than the transaction cost, environmental assistance is rational also when self-interest is assumed to be the sole determinant of state behavior.

It should be noted, however, that environmental gains are in most cases not purely collective, but rather a mix of private and collective benefits. The point here, however, is only to establish the notion of a welfare loss caused by the rule of the least ambitious denominator. If the countries are restricted to the lowest denominator rule, as is the case when unanimous consent is required, the country with the highest marginal cost (i.e. the Baltic State) determines the level of ambition and causes potential welfare losses from unrealized cooperation.

It should be noticed that the lowest denominator rule does not cause any welfare losses when only domestically financed environmental investments are considered. The adopted recommendations by HELCOM set the lower limit of ambition, but any country may adopt regulations that are more stringent. This is also the case in practice, where for example the more affluent Nordic countries in many cases have stricter regulations than stipulated by the recommendations.

In conclusion, provided that the collective part of the environmental benefits is larger than zero and the marginal cost of national investments in the Baltic country is lower than in Sweden, a transfer of resources from Sweden to a Baltic country might be rational even with no assumption of altruistic preferences. This conclusion might, however, be affected by the strategies chosen by other actors. Depending on the prediction of the behavior by the other actors, Sweden might choose not to implement the investment, and instead try to take a free-ride. The next issue to address is thus whether there is also a game theoretic equilibrium where Sweden provides assistance for environmental investments in a Baltic country. We have shown that there is a potential for joint gains, but its realization ultimately depends on the existence of an equilibrium, where Sweden chooses to provide assistance.

7. A NON-FORMAL BILATERAL GAME ON ENVIRONMENTAL ASSISTANCE

A common way to represent games in international relations has been to use simple games in normal (matrix) form. Although the simplicity of a model always has to be balanced against its explanatory power, the use of games in a normal form may be connected with considerable analytical risks (Stone 1999). The extensive (game tree) form is a more fundamental mode of representation than the normal form, since all extensive form games can be converted into normal form, while the opposite is not always possible (Morrow 1994). It can certainly be helpful to transform games to normal form (e.g. when searching for a complete list of Nash equilibria), but when using *only* normal forms, important aspects of the actual choice situation may be lost. Sequential moves, for example, cannot be captured in a satisfactory way in normal form representations. As the sequence of moves is of the utmost importance in many kinds of applications in international relations, the use of extensive representations is often necessary. Furthermore, when using only normal form games, one deprives oneself of the opportunity to use some of the more sophisticated refinements in game theory, which have been developed recently (e.g. subgame perfection and sequential equilibria).[37]

In this study, only extensive form games will be used, primarily because the sequence of moves here is of considerable importance. The games presented here, however, are extremely simple, and should be perceived as mere supplements to the verbal presentation, rather than as analytical instruments in themselves. The basic assumptions underlying traditional game theory methodology will be adhered to, but a substantial effort is made to reduce formalization. In this way, careful theorizing can be combined with high accessibility. It is not denied that formalization generally increases precision, but since this precision is often bought at a considerable cost in terms of number of readers, the formalization commonly used in mainstream game theory will be significantly relaxed here.

It should be noted that formalizations are not intrinsically more valuable than verbal presentations, as ways to suggest reasonable arguments. In principle, there is nothing that can be said in terms of equations that cannot be said in words (Walt 1999). The advantage with using a verbal argumentation is primarily that reason and discursive knowledge on the subject can be more easily adhered to then when only

---

[37] Although the so called Kuhn-Zermelo theorem stating that every finite n-person game of perfect and complete information has a unique subgame-perfect equilibrium in pure strategies, provided that no player is indifferent between any of his/her strategies, was formulated in 1913, it is only recently that this concept has been more widely used by political scientists (Zermelo 1913). The concept of sequential equilibria was developed by Kreps and Wilson in the early eighties, as a way to find solutions based on consistent beliefs and strategies (according to Bayes theorem) (Wilson 1982). With this concept, Nash equilibria that seem obviously unreasonable can be rejected by a coherent mechanism.

pure logic is applied. This does not mean, however, that careful verbal argumentation is always preferable to formalization. At early stages in an exploration of a particular topic, the most productive method is usually to try to find interesting hypotheses and to make informed statements based on reason and knowledge of the topic. Then comes the time for formalization, when the suggestions made earlier have to be corroborated in terms of logical consistency. Since the issues covered in this book have not yet been extensively explored, though, the choice of verbal argumentation seems sensible.

The game theoretical method has been criticized for lacking ways to model dynamic behavior. P. T. Hopmann, for example, states "...game theory is a static model that focuses almost exclusively on the initial relationship between initial conditions and outcomes, with almost no attention being paid to the process of bargaining" (Hopmann 1996:51). Although this is not true, since there have been numerous studies where for example the game is iterated, or models in extensive form have been used, thereby introducing a time dimension and incentives for learning (e.g. Bayesian approaches) and adaptation (e.g. evolutionary game theory), it is true that the achievements in terms of coherent models with a unique or small number of equilibria have been rather limited.[38]

A quite common critique against game theory is that it is not well suited to handling dynamic processes, because its deductions are tautological. When labeling a model as tautological, the implication seems to be that no new knowledge is presented, that the same thing is said twice, but in different words. This, however, is not how a tautology should be looked upon in scientific methodology. In ordinary language, a tautology implies something superfluous. In science, though, a good tautology is the essence of the most prominent methodology, the hypothetical-deductive method. By stating the same thing in different words, understanding can be considerably increased. In fact, all coherent scientific models and theories that abide by the laws of logic are tautological in the sense that they are only deductions from the original assumptions. All their aspects are latent consequences of the assumptions, and the scientific task is to make these latent consequences explicit. This has been succinctly put by Emerson M. S. Niou et. al.:

> Rationality is not a simple elementary concept in our paradigm. It not only embodies a set of axioms about preferences (that they are complete and transitive), but, in addition, much of what game theorists do can be interpreted as the discovery of the meaning of this word in different contexts. (Niou et. al. 1989: 19)

Due to the difficulties on the one hand to find a small set of equilibria in iterated games, and on the other hand not to have an overly complex extensive game, the model presented here will be assumed not to be iterated. The well-known *folk theorem* states that almost any outcome above the minmax values is compatible with a

---

[38] The Bayesian approach has also been criticized, but not so much on its general validity as on how exactly the updating of information should be formalized and interpreted. See for example Bacchus (1990), Nozick (1995), and Thomson (1996).

game iterated indefinitely (Kreps 1990 Ch. 14.2). As has been emphasized by Eric Rasmusen, "… the lack of a unique equilibrium is a major problem in game theory." (Rasmusen 1997:16) The primary reason in this study for not using an iterated games approach is that the actual interactions differ considerably between different projects. It does therefore not seem reasonable to model a situation where the assumption is that identical games are being played a number of times, a so-called *supergame*. It is, however, probably true that the actors become more experienced the more they are involved with projects that have been jointly implemented. Although the interactions over potential projects will differ considerably, the actors probably become somewhat more informed about each other's utility functions as well as other factors influencing the strategic situation over time. This process of learning and reputation-building will be discussed further in Chapter seven.[39]

The game described below is intended to model the typical case. Although there might be particular projects, which do not comply with this simple model, it could be expected that most of them to do.[40] The alternative to this approach would be to include all the projects in a single, dynamic game. This, however, would result in an extremely complex game where the possible outcomes would consist of any mix of cooperative and uncooperative behavior in the full set of projects. Considering that the number of projects assisted by Sweden exceeds 100, this game would most certainly become intractable.

7.1. The incentives facing the actors

When comparing different alternatives for requesting financial assistance from Sweden, the Baltic country has to balance the probability of getting an affirmative response against the predicted national net gain derived from an implemented project. The environmental good a particular project is targeted to realize, may be valued differently by the Baltic country and Sweden, because of diverging opportunity costs and/or because it is a non-perfect collective good. This means that the ranking of different potential projects will not necessarily be identical in Sweden and in the Baltic country. Assuming self-interested preferences and maximization of expected utility, this means that the Baltic country will suggest projects that maximize the product of project utility and probability of approval. The structure of this strategic situation will sometimes be akin to a Prisoners' dilemma. Sweden would like to pick an assistance project in a Baltic country with as much collective good content as possible, while the Baltic State would like to choose one from its own

---

[39] One of the earliest and most influential works on iterated games and the importance of creating a reputation were published by David Kreps et al in 1982. See Kreps (1982) and Wilson (1982). As noticed by, for example, Denrell (2000), the task to formulate a realistic iterated game sequence is often difficult. One problem is that the actors might not draw the same conclusions from past experiences (the history of the game). If they are assumed to do so, the validity of the analysis would most likely decrease.

[40] This approach bears some resemblance to the methodology often called "modeling by example". See for example Rasmusen (1997).

most highly prioritized areas, which might have various proportions of collective good content. When the preferences regarding the most highly ranked project differs between Sweden and the Baltic country, they face diverging incentives. However, a failure to reach an agreement is Pareto-inferior to an agreement. In other cases there might not be a conflict of interest at all, if both actors prefer similar types of projects. It is likely that over time, the actors' information about each other's preferences will increase and the incentives for reputation building will become higher. This will be discussed further in Chapter seven.

It is assumed that the actors have perfect information, i.e. the actors know how the game has been played in earlier phases. But although it is assumed that all actors know *their own* utility function and their valuation of different outcomes, it will not be assumed that this is common knowledge. For simplicity, at this stage it is assumed that Sweden is the sole external financier. In reality, however, it is quite common that more than one external party is engaged in a particular project. In Chapter seven, this restriction will be relaxed, and it will be discussed how this might affect the strategies of the actors.

7.2. The choice situation of the Baltic State

As mentioned in Chapter one, the utility to the Baltic State from a particular support project is determined by three parameters: benefits, costs, and net grants. The gain from the project consists of the sum of gross private and collective benefits. The fact that other countries also gain from collective benefits is not relevant. The Baltic State is only assumed to care about its own environmental achievements. It should be recapitulated that the gains referred to here solely are of an environmental nature. Other potential benefits (e.g. job creation, international contacts etc) are not included. Although this is a simplification, the importance of these other kinds of gains is relatively minor, since the focus here is on comparisons between different projects and unilateral or bilateral financing. In other words, the absolute size of non-environmental gains is not relevant. The simplification is rather that it is assumed that these other criteria are not very important when choosing between different environmental projects, or between types of financing.

The cost of a particular project is the full project expenditure minus net grants. Net grants consist of the external support minus the transaction costs. The transaction factor comprises primarily the costs of reaching an agreement on project implementation. Since reaching the implementation phase is likely to be more complicated when two countries are involved than when a project is undertaken domestically, the transaction costs will be correspondingly higher. The contribution from external sources is what is actually obtained by the Baltic State, not the outlay of the foreign donor. When a domestically financed project is considered, grants are not available, which means that that Baltic State has to cover all project costs.

It should be recapitulated from Chapter one that the values of all variables and parameters are expected values. The only exception is the size of the grant, which is known by the donor country.

Simplifying into a binary decision situation, the Baltic State has to choose between a project completely financed domestically and a project co-financed by Sweden. It is assumed that the budget is given exogenously, which means that the total disbursement is given and the Baltic country can only choose between either a joint or a domestically financed project. In the long run the budget restriction is not fixed. In the short run, which is most relevant here, however, it can be assumed that changes in the allocation to the environmental sector are difficult and costly. It is therefore reasonable to assume that the budget is exogenously given. The first decision-node of the game is shown in Figure 1 below.

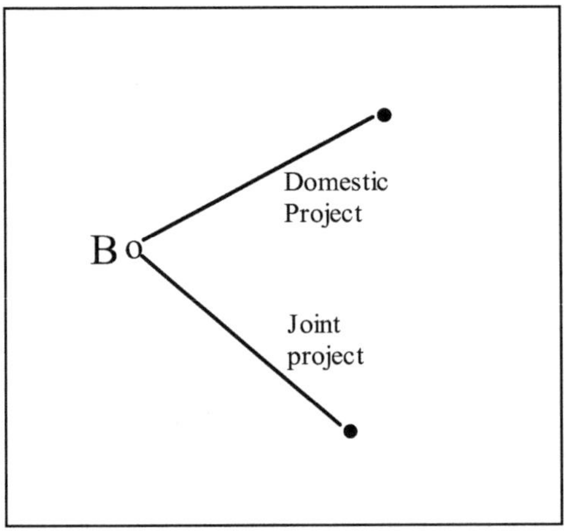

Figure 1. The game begins with the Baltic country (B) choosing between a project financed by domestic means, and one funded with Swedish assistance.

Two things related to the decision situation depicted in Figure 1 above should be emphasized. First, the choice depicted is in the typical case between two different projects. If the project is financed with domestic resources only, the Baltic State is free to choose any project it wishes. As will be elaborated further below, it will in this case choose the most highly prioritized project, according to its agenda of environmental improvement. If the Baltic State, on the other hand, chooses a jointly financed project, the selection process will be the outcome of the interaction between the donor and recipient countries. As also will be elaborated further below, this selection process might very well result in a choice of project that is not optimal to the Baltic State.

Second, since the budget restriction of both countries is assumed to be fixed in the short run, the choice of a domestic project makes it possible for Sweden to spend the resources intended for the joint project on something else. The assumption made here, is that these resources will be spent on an alternative environmental project in Sweden. This seems reasonable, since the fixed budget restriction makes it difficult to make radically different resource allocations. This means that if the Baltic State chooses a domestically financed project it will, in addition to the environmental benefits from this project, also gain from the alternative project implemented in

Sweden, provided that the Swedish environmental project creates not only private gains, but also collective benefits. The latter will benefit both countries identically.

To summarize, the Baltic country has to choose between these two alternatives:

- A domestically financed project, where the environmental gain comprises the sum of private and collective benefits from this project plus the collective benefits from the alternative project implemented in Sweden. The Baltic State has to cover all costs related to the domestic project, while it gets the collective gains from the Swedish project for free.

- A jointly financed project, which is probably less attractive from a purely environmental point of view, since the project selection process was not entirely free. On the other hand, part of the cost is covered by the net foreign grant.

7.3. The Swedish choice situation

The choice of the Baltic country becomes known by Sweden when the project application is received or when Sweden is informed in some other way. In other words, when Sweden is to make its move, it has complete information about the history of the game.

The relevant choice for Sweden to make when it has obtained the request from the Baltic State is to compare the predicted net gain from an agreement to finance the project with the envisaged utility from the most favorable, alternative environmental project. Most likely, this alternative project will be one implemented in Sweden. It should be noticed, that this does not mean that it is assumed that the actual selection process by the Swedish authorities will be between these alternatives. Rather, what is assumed is that these are the options available to Sweden, in terms of policy or program formulation. The projects to be approved are selected by officials in the public administration, and a large number of comparably minor factors may influence the choice of particular projects or assignments, but the guidelines for this selection process are ultimately given by the Swedish government and by the parliament. The situational factors may be important in specific cases, but that does not contradict the assumption that the model captures the most essential features in a more general sense of the portrayed decision situation.

The environmental benefit for Sweden from financing a project in a Baltic country equals its collective good share, since all private benefits by definition are accrued by the Baltic State. Typically, this means that the objective gain from such a project is smaller for Sweden than for the Baltic State, since the latter gains from private as well as collective benefits, while Sweden only gains from private returns. What matters, however, is not objective gain, but subjective valuation. It is assumed that Sweden cares little about the difference in gain between the two countries. Instead, the relevant comparison is collective benefits from co-financing a project in the Baltic State and collective and private gains from an alternative project implemented in Sweden, both in relation to their respective cost. The cost related to co-financing a Baltic project is the grant (including transaction costs). If Sweden were to choose a domestic project instead, it has to cover all costs. It should be noted, that since it is assumed that the marginal return from environmental investments is assumed to be decreasing, the collective benefit from a co-financed Baltic project could be quite significant from a Swedish perspective, due to the fact that the marginal return from Swedish projects is comparably small. This notion will be further elaborated later on.

If Sweden chooses not to take part in the joint project, it is assumed that the Baltic country will implement some other environmental project, financed with domestic resources. This means that if this project creates any collective benefits, Sweden will share these benefits with the Baltic State.

To summarize, Sweden may choose between the following alternatives:

- To agree with the suggested joint project. The gain will then be collective benefits from this project, and the costs consist of the grant (including transaction costs).

- To choose an alternative domestic project, where the gain will be private and collective benefits. The cost will then equal total project costs. In addition to this, Sweden will get the collective benefits from the alternative project implemented in the Baltic State for free.

In Figure 2 below, the full game is shown, where both actors' available choices are depicted.

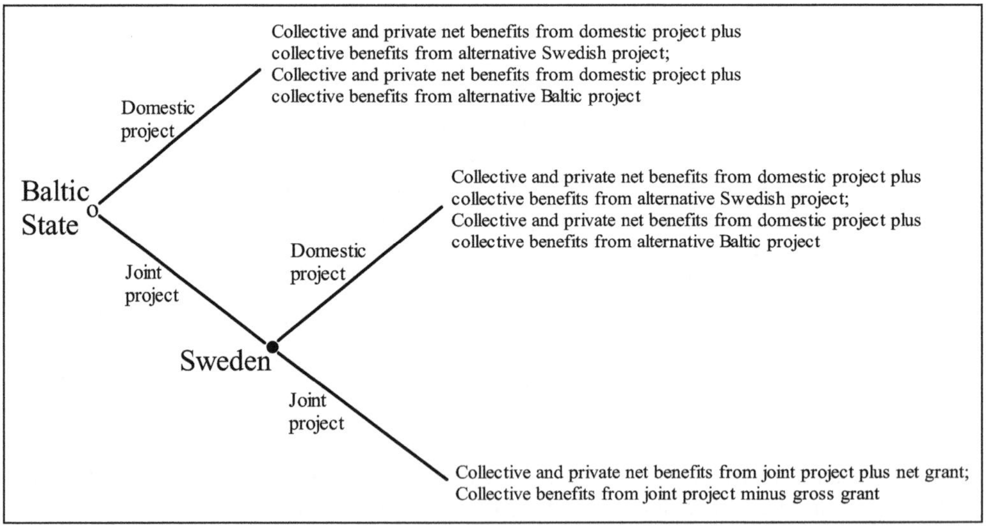

Figure 2. The full game between a Baltic country and Sweden whether to propose a co-financed project or to finance projects purely with domestic means.

Figure 2 summarizes the decision situations and possible outcomes for both donor country (Sweden) and recipient country (Baltic State). First of all, it should be noted that for a joint project to be implemented, both actors have to agree on this. This means that both actors must choose "Joint project" at their respective choice nodes in the game depicted above. At the end of each branch (far right), the outcomes resulting from the chosen strategies are shown. The first player's (Baltic State) is given first, and the last player's (Sweden) is given last. "Net benefits", it should be recapitulated, refer to gross benefits minus gross costs. Similarly, "net grant" refers to gross grant minus transaction costs.

7.4. Equilibria of the game

Using the method of backwards induction, it is clear that Sweden will choose to agree to finance a joint project provided that this gives a higher payoff than to choose a domestic one.[41] This means that for Sweden to choose to finance the Baltic project, the gains derived from *this* project have to be larger than the sum of the net benefit (collective and private) derived from the best alternative Swedish project, the collective part of the project implemented by the Baltic country, and the opportunity cost of the provided grant. The direct relation between the collective good part of the assisted project and the size of the grant should be noted. The larger the contribution, the bigger the collective good part has to be, in order to sustain the cooperative equilibrium. The size of the grant is the only term directly under the control of Sweden. In order to maximize utility, the model predicts that Sweden will only provide the minimum grant necessary in order to make certain that the assisted program will be implemented, other things given. This means that the proportion of total costs covered by Sweden may vary depending on the type of project. Projects with large collective benefits could make it rational to provide comparably sizable grants, if this is necessary to make the Baltic country agree to select a particular project. In other cases, collective benefits are too small to be compatible with the minor grants also on offer. As will be discussed in the final chapter, however, in an iterated game sequence it might be preferable to choose a general proportion of cost coverage. This will for example decrease transaction costs, increase the level of information and give reputational effects.

Moving back one step in the game tree, and if we assume that the inequality described above is satisfied (Sweden will, if it gets the opportunity, accept the Baltic project proposal), we see that in order to make the Baltic country prefer to make the proposal in the first place, net benefits from the assisted project (including the element of net grants) has to be larger than the sum of the net benefit from a domestic project plus the collective benefit from an alternative Swedish project. If this holds,

[41] Backwards induction is a method to find equilibria in extensive games. One starts at the last decision node to see how the actor prefers to move at this node. From this point, one then goes back to the previous decision node and continues to do this until the first node is reached. The complete path to each equilibria can in this way be found.

the Baltic country will choose the project with Swedish assistance rather than an alternative one financed purely by local means. In order to maximize utility, the Baltic State will pick a project with as high benefits (collective and private) as possible in relation to total cost net of net external assistance, with the restriction that the collective good content must be sufficiently high to ensure Swedish assistance.

When the two inequalities verbally described above hold, we have an equilibrium where the project with Swedish assistance will be proposed by the Baltic country, and accepted by Sweden. If one or both do not hold, we will have equilibria where both parts choose to implement projects financed purely by domestic means. It should be recapitulated that in this model, the choice between a domestic and an assisted project is an either-or selection. Although the game may be repeated – a new project proposal is put forward – in each stage there is an alternative *not* chosen. Consequently, if an assisted project is chosen, a domestic one will not be implemented. Since the Baltic country is generally required to cover part of the costs by local means, less money will be available for domestic projects for every assisted venture implemented.

Summing up on the potential of reaching cooperative equilibria, joint projects are preferred when both actors subjectively think that this is in their best interest. This conclusion is, of course, in itself quite empty. But the simple model given above provides us with theoretical concepts useful for increased understanding of actual decision situations. Emphasizing the importance of separating private and collective environmental benefits, net and gross grants, and transaction costs facilitate the interpretation of what actually is in the actor's best interest. In order to specify predictions of actual equilibria, it would be necessary to replace each parameter and variable with empirical data from each and every project. It would furthermore be necessary to address not only these cases when joint action have been agreed upon, but also all renounced project suggestions. Clearly, this would be a huge research task. However, since the focus in this study is not on point predictions, but rather what would be expected to happen at the margin, such a comprehensive undertaking is not necessary. What is sought is not a prediction of what particular projects that should be expected to be acceptable to both actors, but hypotheses related to in what direction expected outcomes would be. In what way is the collective good content relevant for a description of the actors diverging incentives? What alternatives to a joint project are available? How might the preferences over these alternatives differ between the actors? Such questions can be addressed with this model, and will be further elaborated below. The focus will be on the equilibrium where the Baltic country proposes a project, and Sweden accepts it, since it is this equilibrium that is compatible with observations of Swedish assistance.

7.5. Incentives at the margin

According to the model, the size of the benefit from an environmental project depends partly on where it is located. The country where the project is implemented gains from both collective and private benefits, while the other, by definition only gains from collective benefits. In a particular project, the size of the collective benefit is identical to both parties. This means that in all cases where the private benefit exceeds zero, the country where the project is implemented will reap the largest benefit. But it also means, that Sweden will only care about the collective part of the benefits. Since private benefits are significant in many kinds of environmental issue-areas, it would be expected that the average collective good content is higher among projects with Swedish assistance than in the set of projects financed exclusively by domestic means. This is expected not only because Sweden only benefits from the collective parts, but also since Sweden knows that the Baltic country knows that projects with a low collective good content will have less chance of being financed by Sweden. The optimal set of environmental projects from the perspective of the Baltic State would thus include items with high as well as low collective good content, although a bias towards projects with comparably high private content is expected, while the corresponding set from a Swedish standpoint would primarily include projects in the former category. The first hypothesis is thus stated as follows.

Hypothesis 1
The collective good content is higher in the set of assisted projects than in the set of projects financed exclusively by domestic resources.

Regarding the grant element, the most important ignored factor is probably domestic socioeconomic considerations of different kinds, which are all external to the model. Swedish governments have tended to concentrate on areas where technological know-how is high and physical resources are readily available. By doing all procurement in Sweden, the project selection process has been influenced by factors such as regional development ambitions and support of particular industrial sectors. Since the practice of tying Swedish support decreases its value, the Baltic countries' preferences have to some extent been in conflict with Swedish interests to combine foreign assistance with domestic considerations. Since these factors are not an integral part of the model, though, they will only be briefly discussed in Chapter seven.

The size of the Swedish contribution is one of the most central factors in the model. The incentives facing the actors here are strictly opposed, where the Baltic country would, ceteris paribus, like to have as large part as possible financed by Sweden, while Sweden would prefer, ceteris paribus, to pay as little as possible. If we combine these opposed tendencies with the assumption that the alternative cost for an environmental project is higher in the Baltic country, as described at the beginning of this chapter, there seem to be opportunities for joint gains when implementing assisted projects. However, as can be seen from Figure 2.2, there is a difference between gross and net grants. The value sent from Sweden becomes somewhat smaller before it is received by the Baltic State, due to transaction costs. This means that Swedish environmental assistance is not rational all the way up to the situation where marginal benefits from environmental investments are equal in the two countries. From this model, it would rather be expected that a motivation to provide assistance exists, as long as there is a difference in marginal *net* benefits between investments in the two countries. Since transaction costs are incorporated in the net benefit parameter, assistance is only rational as long as there is a difference in net benefits. In other words, assistance is rational as long as the difference in benefits is larger than the associated transaction costs. It should also be emphasized that the stated potential of mutual gain is purely conceptual. Whether or not there is such a potential in a particular case depends on the character of the project (i.e. if the collective good content is sufficiently large).

Marginal benefits from environmental investments are very difficult to estimate, and almost impossible to measure with any high degree of precision. This might seem to detract analytic value from the model described so far. This is not the case, however, since it can be reasonably assumed, at least as a theoretical estimation, that marginal benefits in different public sectors are identical. When net benefits from investing in sectors other than the environmental are larger, public resources should, theoretically, flow towards these other sectors. Ultimately, this results in equalization of net marginal benefits in all sectors. If the simplification then is made that the percentage of spending is equal in both countries, it follows that a difference in national income between the two countries, implies a difference in marginal benefits (transaction costs excluded) from environmental investments. Thus, provision of assistance is rational as long as there is a difference in marginal net environmental benefits, and this difference is directly related to differences in national incomes. Admittedly, the assumption that relative public spending is equal in both countries is not very realistic. However, since data on relative public spending is readily available, this simplification does not detract analytical value from the model. As long as relative public spending is lower in the Baltic State than in Sweden, environmental support is rational, also when differences in national incomes are held constant.

The situation captured so far is static. The intent has hitherto been to show the underlying incentives driving a process already well underway. It has been shown that there is a potential for mutual gains when the difference in national income is larger than transaction costs related to joint implementation of environmental projects and differences in relative public spending are controlled. So far, however, it has been assumed that the recipient country, the Baltic State, has been fully sovereign in terms of allocating public spending. When this assumption is relaxed, it is possible that the equilibrium described informally above will not occur. The reason for this is that the Baltic State is no longer fully sovereign in terms of allocating public spending. Sweden, as well as most other donor countries, requires each project to be partly financed with local resources. This means that Sweden might very well have incentives to provide further investments in the environmental sector (marginal net benefits are larger from investments in the Baltic State than in Sweden), while to secure general equilibrium in the economy, further investments should rather be implemented in other sectors. Unfortunately, this means that, over time, differences in national income will be a more biased measurement of the strength of the incentives to provide assistance. When the environmental investments in the Baltic State become "too high" in relation to other sectors, differences in national income overestimates the marginal return from environmental investments. In other words, we are back in a situation where marginal net returns from the respective countries must be directly compared in order to show the magnitude of potential mutual gains. Over time, as more joint projects are undertaken, the potential for further mutual gains will decrease, since the marginal net returns decline. This holds under the assumption that the benefits emanating from a particular project have a reasonable durability (i.e. an implemented project continues to be a benefit during the considered time frame). Ultimately and provided that no external disturbances occur, the difference in national incomes is not changed, and relative public spending is unaltered, net returns from domestic projects in Sweden and joint projects in the Baltic State will tend to converge. This implies the second hypothesis.

> *Hypothesis 2*
> The number of initiated projects will be high in the early phases of the interaction between Sweden and the Baltic country, and thereafter stabilize and only vary with, for example, relative changes in national incomes, relative public spending, or external disturbances.

What is particularly interesting about the potential from joint financing described above, is that domestic resources in the Baltic country will tend to flow from environmental issue-areas not particularly collective in their nature. Since the Baltic country generally is required to finance part of the project with local means,[42] resources will flow to projects where external financiers cover a large part of the project costs, which means that other projects, which have not been prioritized by these external parties, will get less. This effect will be smaller when additional parties other than Sweden cover part of the project cost, but it will probably not disappear, since most other bilateral assistance providers can also be assumed to be primarily interested in projects with high collective good content. Furthermore, in order to increase local participation and cost-consciousness, most providers of external resources will not finance a project completely, even if they could do so very well in concert. The requirement of some level of domestic financing is chosen more because of a perceived need to curb strategic action on the part of the recipient country, than because of pure financial reasons. If the recipient country was not required to finance any part of a particular project, it would be possible for them to strategically choose to domestically finance only projects in issue-areas with low collective good content. Knowing that bilateral donors would prefer to finance projects in issue-areas with high collective good content, it would be natural from the perspective of the recipient country to let these providers of assistance cover as much of these costs as possible, and use domestic resources for investments in areas of little interest to donors.[43]

It should be noted, however, that the recipient country is better off with assistance than without, even though the bias caused by the donors makes the over-all utility lower than would have been the case if the recipient country had been allowed to invest in issue-areas solely according to its own discretion. If this were not the

---

[42] See for example Regeringsbeslut UD20000/276/AF.

[43] This general possibility to act strategically to maximize external support has been noticed by for example Gil & Folmer (1998).

case, the Baltic State would simply choose not to accept further joint projects.[44]

Finally, it should be noticed, that the duration of an environmental investment could be assumed to be lower, when external financing was pivotal to its implementation. As it has been an explicit strategy by Sweden not to support projects over their complete lifetime, but rather to reduce dependence on foreign sources in the long run, the Baltic country will have to cover more and more of its costs. As the project by definition was not sufficiently high on the priority list to be selected without external assistance, there may be a point where the continuance of the project is no longer optimal. The location of this point will depend mainly on the difference in utility between this project and alternative ones, and on the relation between sunk and marginal costs. In this sense, external financiers might decrease the duration by initially covering large parts of the investment costs.

> *Hypothesis 3*
> Externally financed environmental projects will tend to withdraw domestic resources from nationally prioritized areas where the collective content is low.

Before the model presented here is validated against data relating to the Swedish environmental support portfolio, the evolution of Swedish development assistance and international, environmental cooperation, the political consensus on the profile of the CEE assistance, the administration of the assistance, and the contemporaneous environmental preferences of the Baltic States will be described in the following chapters. By doing this, not only will the validity of the model be enhanced, but an extensive orientation of the process of interaction between Sweden and the Baltic States in the area of environmental protection will also be provided.

---

[44] However, a particular joint project with negative utility to the Baltic State could in fact be accepted, *if* it is seen as a precondition for further cooperation, and the expected, discounted value of the latter is larger than the loss of utility related to the particular project.

CHAPTER 3

ROOTS OF CONTEMPORARY SWEDISH ENVIRONMENTAL ASSISTANCE

1. VALIDATING ASSUMPTIONS ON PREFERENCES

In this chapter the historical background to the Swedish assistance policy vis-à-vis the Baltic States in the 1991-1996 period is presented. This background primarily serves as a basis for the formulation of Swedish preferences in this issue-area, and thus lays the ground for the motivation of the assumed utility function used in the model. As it is assumed that the policy pursued by Sweden vis-à-vis the Baltic States in the area of environmental cooperation is primarily guided by self-interest, it is highly relevant to describe the historical tendency for making a joint gains approach an integral part of the over-all foreign aid policy of Sweden. It will be described how the Swedish foreign aid became dichotomized with one part characterized by altruism and the other part primarily targeted towards mutual gains. The latter part will be of particular importance when trying to understand the CEE assistance policy adopted later on.

Some authors have seen the formulation of actor preferences as one of the weakest parts of most rational choice approaches in political science (Udehn 1996). In economics, the construction of individual utility functions in most cases is not problematic (except for in e.g. welfare economics). It is generally assumed that the individual prefers a lower price for a good of some kind, than a higher, ceteris paribus. In political science, however, most of the time it is a lot harder to make reasonable assumptions about the preference order of the actors. The lack of expressed valuation of the particular good is what makes the analysis much more shaky, the lack of a market price.

It seems intuitively true, that when defining the preference order of the Swedish government concerning cooperation activities with the Baltic States, as much information as possible should be used. Fritz W. Scharpf has convincingly argued, that:

> If we cannot have empirically validated theories that combine broad scope and empirical specificity, we nevertheless are not compelled to face each new case as if we had no prior expectations about how the world works. Instead, we should be able to explicate what the authors of "good" case studies always have in the back of their mind: a "framework" that organizes our prior (scientific and pre-scientific) knowledge about what to expect in the province of the world that is of interest to us... (Scharpf 1997:29)

Although Scharpf primarily refers to the general use of different theories, the line of argument is no less relevant in the context of formulation of basal assumptions such as the utility functions of the actors. The historical and institutional background of the Swedish CEE assistance program will thus be explored in order to make the theoretical assumptions more compelling.

In the analysis of the 1991-1996 environmental support a comparative static approach will be used, where it is assumed that the preferences of the Swedish government are exogenously given. Or in other words, they are not allowed to change during the time period covered by the analysis. If this were allowed, it would not be possible to distinguish between changes in preferences (what the government *wants* to do) and the structural setting (what it *can* do). This does not mean, however, that preferences cannot change over a longer time period. In fact, in this chapter, one of the main themes will be the changes in Swedish foreign aid policy since the 1960s.

Throughout this study, the analytical distinction between preferences and structure will be emphasized, what actors would like to do and what they can do (i.e. the difference in relative costs connected with alternative actions). As will be described below, the relative costs of choosing different strategies in this issue-area have been influenced by a number of factors. Examples of such exogenous factors are: increased international interdependence, the performance of the economy, the strategies chosen by other countries concerning foreign policy, new international institutions etc. Such changes in structure do not analytically influence what the Swedish government *would like* to do, but certainly influence the costs associated with different strategies.

2. DICHOTOMIZATION OF SWEDISH FOREIGN AID

Foreign policy is always a result of a merging process where different interests are molded into a more or less coherent whole. The policy adopted is not determined by a single actor, but is rather an outcome of interactions between different actors. Thus, the interests of the relevant sub-national actors must be traced. By doing this, important clues to the rationale behind specific foreign policy outcomes can be found. Furthermore, it is necessary to find the relevant structures restricting the individual's freedom of action. The outcome of this process may subsequently be used as an input into the model, as the rationale behind the formulation of actor preferences. Hence, the aim in this chapter is to map the influences on contemporary Swedish environmental assistance from domestic and historical processes. The motivation behind this is *not* to give an exhaustive account of this process, but rather to substantiate the formulation of actor preferences. These preferences – which are purely theoretical constructions since in reality they do not belong to any individual actor, but are a reflection of diverging domestic interests – need to be very carefully defined. It is not à priori necessarily true that the state "actor" should act in an individually rational way, since this depends on the domestic distribution of interests.

To trace the evolution of the Swedish preferences in this issue area, earlier Swedish development aid and environmental policy is described. The former will be approached focusing especially on signs of non-altruistic intentions, while transnational and regional considerations will be given the main attention in the issue-area of environmental policy. The aid policy of Sweden will, however, be addressed somewhat more extensively than the environmental policy. It is reasonable to assume that a higher degree of diversity in terms of different interests will be found there. Environmental issues may also involve competing interests, but the primary clash is not likely to be found between different opinions on self-interest and altruism, but rather on how to optimize environmental enhancement within the constraints of a given budget.

More specifically, it will be shown that there have been two main trends evolving over time: an increased awareness of the transboundary characteristics of many environmental disturbances and a split of Swedish foreign aid into one branch emphasizing altruistic aid to LDCs and one branch focusing on mutual gains in the contacts with middle-income countries. The increasingly apparent dichotomization of Swedish foreign aid was caused by several factors. Two were of special importance; (*i*) an ideological drift towards focusing on aid efficiency and the significance of domestic policies adopted in the recipient countries, and (*ii*) economic recessions in Sweden, which increased the pressure from business and labor interests on the composition of the foreign aid portfolio. The ideological drift was not something unique for Sweden, but rather a change of political direction in most OECD-countries. The political radicalization of the 1970s was replaced by pro-market and liberal sentiments in most Western countries in the following decade. With this emphasis on (free) trade, notions of joint gains also in contacts with developing countries increasingly expanded political legitimacy.

The downturn in the Swedish economy in the 1970s and early 1980s placed additional pressure on the foreign aid budget. There were demands that greater attention be given to how domestic business and labor interests were to be accommodated within the foreign aid context. The outcome of this pressure was an increased use of tied aid and later the concept of return flows.

The combined consequence of the ideological reorientation and the pressure exerted by business and labor interests was a rupture in the earlier quite homogenous composition of Swedish foreign aid. The eradication of poverty was no longer the only major goal. In addition, Swedish foreign aid should now also promote a wider area of inter-state cooperation as well as, for example, strengthening democracy and human rights. It will be shown below, that this change, together with the perceived increase of the relevance of international cooperation with regard to combating transboundary environmental disturbances, were important factors behind the institutionalization of Swedish environmental support to the Baltic States in the 1990s. The political changes in the former socialist countries made it possible for Sweden to take on a more active role, while the domestic alterations provided the impetus for it.

3. THE PRACTICE OF TYING AID

The practice of using *tied aid* (or, as rephrased in the 1980s, *return flows*) is of particular interest in the context of this study. Tied aid can be defined as foreign assistance where the recipient country's choice of supplier must follow certain rules defined by the donor. The references to "the Swedish resource-base", "specific knowledge of Swedish firms" etc are legio in most policy documents of development assistance.

The tying of aid *can* in some cases be defended on purely economic grounds, but in most cases it results in attachments of different political interests being put on aid projects.[45] These attachments are not connected with the strict objectives of the aid programs. *If* a certain product or service can be produced in the donor country at a marginal cost considerably lower than the world market price, *and* the recipient country (a) would not have bought the goods or services to the market price and (b) is not allowed to resell it to some other buyer, a case can be made in favor of using this particular item as tied rather than monetary aid. The real cost to the donor is the marginal production cost of the good. This is the relevant cost to the donor since the market price is by assumption not affected, and consequently neither is company profit. By giving the goods instead of the money, the donor can give more for the same cost. The same phenomenon is quite often seen in companies, when employees are offered fringe benefits (company products) at a reduced price, if they promise not to resell the goods. As long as the price is not lower than the marginal production cost, the company will not suffer from this policy.

The value to the recipient country of tied aid is somewhat more complicated. Since no figures on willingness to pay for the product in question is available, the valuation of it cannot be measured. *If* it is true that the country values it higher than the production cost (in other words; if it would rather have the product than the amount of money corresponding to its marginal production cost), then a purely economic case can be made for tying the aid to this particular item, rather than offering the recipient completely untied aid.

A more common situation, however, is when donor country's domestic considerations are allowed to affect the orientation of its assistance policy. Aid is not tied because it makes it more economically efficient, but because this policy benefits the donor in some way or another. As we will see below, demands from business – as well as from union interests - may exert strong pressures on the composition of aid programs. There are often strong incentives for the company, industry, or union to make it more directly beneficial to themselves. There may also be political incentives to support interest group considerations on how the aid policy should be formulated. These political incentives *may*, from a welfare point of view, result in a legitimate favoring of particular business or labor interests if, by doing so, positive externalities outweigh the efficiency losses (e.g. directing the procurement to a specific region one wishes to support). The more general point, however, is that these efficiency losses are *not* legitimate from an economic welfare perspective. The specific company or union may of course benefit - that is why they lobby for this particular aid program - but the same is generally not true from a country perspective, taking into consideration the utility to the recipient.

Tied aid is very susceptible to political biases. It can readily be shown to the

---

[45] Interestingly, according to M. Radetzki "The efficiency losses due to the tying of the assistance flows are denied by Swedish donor agencies." (Radetzki 1994:9) It should be noticed, though, that even if the support to the Baltic States is within the same program, the statement by Radetzki concerns only Poland.

public that a particular aid package benefits specific business or union interests. This is the most important explanation of the use of tied aid and the weight put on return flows. On economic efficiency grounds, however, the potential losses from selective business support may be found in any introduction textbook on economics. But as a rule these losses are only possible to discern at an abstract and long-term level. Therefore, these arguments are not as politically powerful as selective industrial support policies are.

4. EARLY SWEDISH FOREIGN AID

Foreign policy processes always have a history. It is rarely evident where to cut off the threads to earlier events.[46] In the case of Sweden, the birth of a coherent foreign aid policy has often been dated to 1962. In this year, the government budget bill proposed a marked increase in foreign aid, and also made a clear statement about the moral obligation to support poor countries and how this should be done. Therefore, this event has been chosen as the starting point when tracing the roots of Swedish foreign aid policy.

In the sphere of international environmental cooperation, more extensive activities began in the early 1970s.[47] In 1970, the Organization for Economic Cooperation and Development (OECD) established an environmental committee, and in the same year the UN Economic Commission for Europe and North America (ECE) set up a formal group with the purpose of discussing transboundary air pollution (SOU 1990:88).

The UN Conference on the Human Environment in Stockholm, 1972, was in many ways both a confirmation of the increasing concern for transboundary pollution and a starting point for institutionalized, international environmental cooperation. Of particular interest was the initiative by the Finnish delegation to draft a regional convention on environmental cooperation in the Baltic Sea region. This convention was signed and an interim secretariat was established in 1974. This is the primary reason why 1972 has been chosen as the starting-point for the search for the roots of Swedish environmental aid to the Baltic States in the field of international, environmental cooperation.

4.1. International context

The formulation of a country's foreign aid policy is affected by influences from other donor countries' policy choices. The interplay between the actors, as well as processes taking place in other issue-areas such as national security, induce changes in the formulation of foreign assistance. Some of these forces affect different coun-

---

[46] Nobel Prize laureate Douglass C. North has elaborated the issue of path dependence eloquently. See for example North (1990).

[47] The earlier international conventions in this area were almost exclusively aimed at the protection of animal and vegetation species (UNEP register of international conventions).

tries in a similar fashion. Foreign aid trends become discernible. The most obvious cases where similar trends can be observed are those originating from large and profound changes in the international, political system. The two most substantial changes in the international system affecting the formulation of foreign aid policies in the late 1950s were the cold war between East and West and the dismantling of the European colonies.

The relations between the United States and the Soviet Union were quite strained by the end of the fifties. From its relatively isolationistic profile before the Second World War, the United States was now clearly becoming a superpower with global power projection capabilities. Compared to 1938, military expenditure in the US had increased 13-fold in the 1950s. NATO and the Rio Pact were institutionalized by the United States, and in 1955 no less than 450 American bases were active in 36 countries (Lundestad 1985:11). The United States furthermore had a tremendous influence over the world economy. Before the Second World War, the USA had a dominant position, which in the 1950s became even more accentuated, evident in the fact that almost half of the global production at that time was manufactured in the US.

The Soviet Union could not match the American economy in any way. While production had increased significantly in the United States during the war, the opposite was the case in the Soviet Union. On the military scene, however, the Soviet Union continuously decreased the gap during the 1950s. The creation of the Warsaw Pact in 1955 institutionalized the conflict between the two blocs, and by the end of the 1950s almost no region in the world was unaffected by their discord.

There were furthermore strong tensions building up in the North-South dimension. The European colonial powers were losing their grip on their colonies, first in Asia in the 1940s and then also in Latin America and Africa in the following decades. The outcome of this process was often violent conflict as well as incentives to create other forms of relationships between former colonies and the Western states. At the intersection of these dimensions lay the quest for geopolitical spoils and influence, thereby connecting the East-West tension with the North-South traction. The decolonialization process was, in important ways, conditioned by the outcome of the Second World War and the conflict between East and West. After the war the European powers were dethroned from their position as world leaders, and replaced by the United States and the Soviet Union. Neither of these had any notable colonial power history. Although the United States did not refrain from grabbing some colonies (most notably the Philippines), the force with which the USA worked towards liberation of the colonies influenced the decolonialization process considerably. The United States felt morally obliged to defend the right of all nations to be autonomous. The fact that it definitely was in its interest to detach the European countries from their colonies, reinforced this ideological position.

Both the East-West conflict and the decolonialization process influenced the early aid policies adopted by the developed countries in the 1950s and early 1960s. The volume of international aid from the larger Western countries in the early 1960s was increasing. As the former colonies were weak in institutional as well as economic terms, assistance from the Western states was perceived as a prerequisite for development.

In 1960, almost all Official Development Assistance (ODA) was given by the United States and the former colonial powers. Of a total of 4 807 billion $US, the part contributed by the USA, Great Britain, France, Portugal, Belgium and the Netherlands constituted more than 93 percent (Government bill 1962:100).

The United States held a very prominent foreign aid position, providing almost 60 percent of the total assistance from OEEC-countries.[48] None of the non-colonial powers was even beginning to approach the goal of giving one percent of its gross domestic product, as adopted in the UN general assembly in 1961. The contribution from the nowadays comparatively generous Nordic countries was in 1960 very modest, between 0.05 and 0.1 percent of their GNP.[49]

Regarding the destination of the aid, no clear pattern can be discerned. Although the larger part was given to the poorest countries (less than $US 100 in average income), the largest individual recipients were all located in South East Asia (particularly South Korea and India). This geographical area was at that time of great political importance to the United States.[50]

> ... the great aid-giving countries focus their help to a limited number of countries where they have specific political, military or economic interests... A comparably large part of the American assistance has been given to South Korea, South Vietnam and Formosa [Taiwan]. The French aid almost entirely goes to former or present French colonies, while the British help has been primarily directed to less developed regions in the Commonwealth (Government bill 1962:100 p. 107).

As a general rule, most of the assistance was given as tied aid, i.e. the recipient countries were obliged to choose donor countries' equipment and services. In addition to this, the tendency to buy from the donor, even if it was not explicitly required, was quite strong. From the start, there was a clear linkage between foreign aid and the perceived mercantilist need to support domestic export interests.

The allocation of development assistance at this time was, to a considerable extent, focused on bilateral aid. Only about ten percent was given via multilateral organizations and agencies.[51] The more tangible outcome from the work in international fora such as GATT and the special agencies of the United Nations had yet to come, but substantial resources were already being used for technical and fi-

---

[48] The OECD replaced the OEEC in 1961.

[49] The figures for Sweden, Norway and Denmark were in 1962 0.06, 0.1 and 0.1 percent of GNP respectively.

[50] The total aid given to countries with an average income of less than US$ 100 was in 1962 2 325 million US$, in countries between 100 and 200 1 040 million US$ and in countries with an average income exceeding 200 US$ 944 million US$. Of course, these three groups are not fully comparable; in the lowest income group, there are eight countries, in the other two seven. The average amount given in each group is 265 (Std. 274), 130 (std. 148) and 118 (std. 83). As the figures on standard deviation show, the spread within the groups is large. If we remove the two largest receivers (South Korea and India) from the lowest income countries, the three groups receive about the same total amount of Western aid. Source: Government bill 1962:100.

[51] UN document E3556, 1961, International economic assistance to the under-developed countries; statistics of official contributions in 1960.

nancial aid. (Government bill 1962:100)

Almost all multilateral technical assistance in the early 1960s was channeled through the United Nations and its specialized agencies. The largest distributor, measured in amounts disbursed, was the Extended Program for Technical Assistance (EPTA). These development projects were, however, implemented by the UN specialized agencies (i.e. FAO, WHO, ILO and UNESCO). The budget of EPTA was based solely on voluntary donations by member countries. At this time, the UN Special Fund was also evolving as one of the dominant UN actors. The rationale behind this Fund was to facilitate large investments in developing countries, through prefeasibility studies, education of workers and technical staff and applied research.

Among the multilateral suppliers of financial resources, the International Bank for Reconstruction and Development (IBRD) and its "softer" sister organization the International Development Agency (IDA), instituted in 1961, were the most important. Although formally connected to the UN family, the World Bank has always been able to formulate its own policy to a large extent. The institutional culture and code of conduct of IBRD resembles that of a bank rather than an aid organization. A reflection of this is the fact that in most countries, the department of finance, rather then the foreign (aid) department, is the contracting part in the member countries. While the UN system regulations are based upon the principle One member (country) – One vote, the World Bank assigns its member rights according to their shares, which are correlated with amounts put at the bank's disposal. This rule probably makes the institution more efficient in terms of reaching decisions, since fewer actors are able to have their say on allocation schemes, but it also means that most recipient countries of financial resources are unable to influence the size and conditions of provided loans. In this sense, the World Bank institution might be seen as less democratic.

In the first years of the 1960s, the lending of the IBRD amounted to approximately 700 million US dollars, while the corresponding figure for IDA was approximately 180 million US dollars. The lending from the IDA was given under considerably softer agreements. "In reality, the World Bank and the IDA are identical in everything except their financial terms." (Cassen 1987:5) All IDA loans were given for a period of 50 years, and no interest and no mortgages were charged in the first five years.

4.2. Swedish aid becomes institutionalized

Although most Western states have an earlier history of e.g. missionary activities, it was only in the late 1950s and early 1960s that the aid programs became a distinguishable part of their relations with other countries. Swedish official development aid became institutionalized with the creation of a particular agency for this purpose, The Central Committee for Swedish Technical Assistance to Less Developed Areas (CK), in 1952. This committee, however, was given quite modest resources. In its nine years of existence, total expenditure was not more than approximately USD 4 million (Heppling 1986). The real boost for Swedish development aid came in the first years of the 1960s, with the creation of the Agency for International Assistance in 1961 and with the 1962 Government bill on development aid, where sharp increases in levels and ambition were envisioned. The approach adopted in the Government bill was quite idealistic, with an ambition to keep aid activities "clean" from commercial or other non-altruistic considerations. This approach was, however, modified during the decades to come.

The performance of the Swedish economy was strong in the first half of the 1960s. Due to the huge needs for investment in Europe after the war and the fact that Swedish industry remained intact, growth had been impressive in the 1950s and early 1960s. The Social Democratic Party had been in power almost without interruption since the early 1930s, and the Swedish Model was at the beginning of its period of greatness.

The general optimistic mood among the Western countries in the early 1960s about their ability to make the economies of the developing countries "take off" was well represented in Sweden too. What was needed, it was perceived, was only to get the basics going: "...there was a widely held belief that once certain state institutions, basic infrastructure and competence were in place, development would follow." (Utrikesdepartementet 1994:12) In fact, in these hey-days of the Swedish Model, the optimism seems to have been even greater in Sweden than in many other countries:

> [There was a belief] that a middle way had been found to ensure a lasting peace between economic efficiency and democratic values... The prevailing consensus in Sweden promoted, for instance, the rather extraordinary idea that there was a *natural* harmony between growth, redistribution and democracy (Anell 1986).

In Sweden, as in most other Western states, the government apparatus was believed to be the obvious agent of assistance to the developing countries: "Although the missionary and other non-public activities continued more or less in their traditional ways outside of the public aid channel, there was wide political support for the "statist" approach. Not even the Conservatives opposed it". (Utrikesdepartementet 1994b)

The details of the Swedish foreign aid policy were shaped somewhat differently compared with most other Western countries. The large donor countries were, at least indirectly, criticized for either making their aid too conditional upon donor country security policy objectives (e.g. the United States) or exceedingly attached to relations with former colonies (the European ex-colonial powers). The Swedish aid volume was, however, not very impressive (expressed in percentage of GNP) in the early 1960s, compared to that of the larger Western countries.[52] This was partly defended in the 1962 Government bill:

> ...the Swedish assistance efforts have up to now been not insignificantly lower than those of most other OECD-countries. This can, however, essentially be attributed to the comprehensive assistance the large countries and colonial powers are giving those developing countries where they have considerable economic, political or military interests to look after (Government bill 1962:100 p.4).

[52] In 1960, Sweden allocated 0.06 percent of its GNP to ODA (Government bill 1962: 100).

It was explicitly stated, that Sweden had a certain mission to fulfill: To provide the developing countries with assistance contingent neither on national security objectives nor on old colonial ties. The developing countries should not become politically dependent upon foreign aid providers. It was clearly declared that Swedish assistance ought not to be guided by strategic motives (Government bill 1962:100).

The potential use of foreign aid as a political instrument was not completely disregarded. The purpose, however, was the opposite of "*Real Politik*", as Swedish development aid should be used as a way to "...contribute to equalization and greater understanding among nations and thereby further international solidarity and peace." (Government bill 1962:100 p.6)

Although the Swedish government downplayed the importance of the gap between the amount of Swedish aid and that of many of the other OECD-countries, it was nevertheless seen as important to increase the level of Swedish assistance. Sweden voted affirmatively for the resolution adopted in late 1961 by the UN General Assembly, aiming at reaching the goal of giving one percent of the combined GNPs of all developed countries in foreign aid as soon as possible. This resolution was, however, not judicially binding. The affirmative response to the UN resolution was confirmed in a statement by "Statsutskottet" (1961:128), where the one percent target was said to be approved by Sweden, and that its government would actively work for its accomplishment. A parliamentary decision on this matter, though, was not reached until 1968. As a consequence of the UN resolution, the Swedish government felt obliged to at least give a clear signal of moving towards this goal. To this end, it was proposed to increase technical and humanitarian assistance from 42 to 60 million SEK, to make an additional payment of 20 million to the International Development Fund, besides the statutory fee of 10 million, and to reserve 20 million SEK for bilateral, financial assistance. All in all, this was a considerable increase (approximately 150 percent) in the volume of Swedish aid, compared with the previous year.

To facilitate the implementation of this enlargement of the aid program, and to institutionalize a greater Swedish commitment, a committee working with aid-related issues was set up.[53] The committee was convened in 1961, and began working in January 1962.

4.3. Targets of the early Swedish aid program

The focus of early Swedish aid was set on poverty, as the ultimate aim was to "...raise the people's standard of living." (Government bill 1962:100 p.7) The more concrete goal was ambitiously set to:

- Alleviate famines and mass poverty

---

[53] This committee was labeled NIB (Nämnden för internationellt bistånd), and was to be followed by the establishment of SIDA (Swedish International Development Agency) in 1965.

- Eliminate epidemic diseases
- Reduce the levels of child mortality
- Create over-all fairly acceptable standards of living. (Government bill 1962:100 p.8).

These goals may with hindsight seem quite unrealistic. They should, however, probably not be interpreted as targets to be reached at a predicted time in the near future, but rather as statements of where the ultimate intent of the aid resided. What the eventual rationale behind the provision of development aid really was, boiled down to its essentials. The stated goals were accompanied with a quite impressive understanding of the difficulties in most kinds of international assistance. It was clearly appreciated that a prerequisite for economic development was investment in physical, as well as human, capital. But this was not enough. These investments should always be aimed at being a trigger for a self-sustainable and accumulative process of economic development. And this could not be achieved with foreign aid alone, but had to be accompanied by local participation and influence over the design of particular projects. In addition to this, the need for family planning was emphasized. A growing economy could still result in a decreased level of GNP per capita if the population was growing too fast.

The emphasis on the importance of taking the specific cultural and social setting into consideration when implementing aid programs gives a strikingly modern impression of the early Swedish foreign aid policy. Forestalling the later Swedish debate on which regimes to support, the difficulties in defining political and social goals within the aid-programs in any detail were especially pointed out. Only more general policy goals – such as promoting democratization and socio-economic equality – were seen as realistic to formulate. The specific outline of each project or program had to be customized to fit local conditions. This in turn made it important to deepen the knowledge base in the foreign department bureaucracy regarding recipient characteristics. This was taken *ad notam* in the Government bill 1962:100, where considerable increases in the size of the staff working with aid-related issues at the Foreign department were envisioned. (Government bill 1962:100 p. 9)

Before the 1962 Government bill, Swedish governmental aid was almost exclusively distributed through multilateral institutions. Although the Swedish bilateral aid door was slightly opened in the 1962 bill, when technical assistance on a bilateral basis was made a budgetary item of its own, the emphasis was still very much on the multilateral side of governmental aid. An exception to this was the proposal of an initialization of Swedish, bilateral financial aid of approximately 20 million SEK. The principles for allocating this financial assistance, however, were not clearly formulated in the 1962 Government bill. The potential domestic advantages with this kind of aid were mentioned, but so was the risk of economic efficiency losses.

The government saw multilateral aid as preferable to bilateral assistance for several reasons. In general terms it had the advantage, from the perspective of the recipient, of making the donor less visible and less able to influence the specific design of aid projects. Hence, there was less risk for donor-biased aid designs, when using an international institution as the implementing actor. In addition to this, the bureaucracy in the recipient country might have difficulties in coordinating aid flows, when the number of donors was very large. Multilateral aid to some extent circumvents this problem by merging the resource flows into a single, more integrated one. The critique of the large multilateral institutions as being inefficient and too bureaucratic was dismissed in the 1962 bill in quite a casual way:

> [These] ... shortcomings should not make the donor countries decrease their support for the multilateral institutions. They should on the contrary, result in increased support. Active Swedish support in this direction may be one of the most valuable contributions our country can make in international aid.

During the rest of the 1960s, Swedish foreign aid policy continued along very much the same lines. The yearly expansion of the amounts devoted to foreign aid averaged about 25 percent, making the percentage of GNP increase from approximately 0.1 to 0.5 percent by the end of the 1960s (Frühling 1986:29). The UN goal of setting aside one percent of the developed countries GDP for foreign aid purposes played an important role in the Swedish foreign aid discourse. The intention to eventually reach this target was already appreciated in the 1962 Government bill. In fact, the Swedish statement was even more far-reaching than that of the UN, since foreign aid was defined in a considerably more restricted way, including only official grants, long-term credit and contributions to international aid agencies. Although the Parliament did not make the one percent goal a formal statement of intent until 1968, it did serve as a focal point for policy discussions from the early 1960s onwards.[54]

The rapid increase in spending did, however, put strong pressure on the foreign aid bureaucracy. The quite carefully expressed interpretations in the 1962 bill were difficult to keep in line with the spending requirement. The administrative and analytical capacity did not expand at a similar high pace, resulting in a tendency to overspend on expensive physical investments without always choosing the most cost-efficient solutions. Criticism of this tendency to work towards a quantitative target rather than focusing on the effect of the assistance in the recipient country, was going to be a recurrent theme in future policy discussions.

4.4. The increased aid proponents and their adversaries

Swedish foreign aid policy was generally seen as relatively generous, compared to the aid programs of many other countries (Elgström 1992: Chapter 4). At first glance, this may quite easily be attributed to the strong position of the Social Democratic Party (SAP) and its emphasis on solidarity and the building of a welfare state. The notion of solidarity and equity should not halt at the nation's border, according to SAP. There were strong currents in the party advocating further increases in the aid budget. But it is not correct to say that this was the only opinion within the SAP. As a matter of fact, in the early sixties there was a political controversy be-

---

[54] The importance of focal points has been eloquently described in Schelling (1960).

tween the Minister for Development Cooperation, Mrs. Ulla Lindström on the one hand, and the Minister of Finance, Mr. Gunnar Sträng, on the other, which ultimately led to the resignation of Mrs. Lindström. The Minister of Finance seems to have been backed up by the Prime Minister and the Cabinet (Andersson 1986). It has been suggested, that the proponents of dramatically increased levels of Swedish foreign aid formed quite a small group, predominantly consisting of liberal intellectuals (Andersson 1986). The view of the Ministry of Finance was that foreign aid was not particularly popular among ordinary people and it was consequently difficult for the Foreign Department staff to motivate sharp increases in governmental, budget negotiations.

There was, however, a quite strong coalition of foreign aid proponents in the second half of the sixties, consisting of a faction of Social Democrats, the Center Party, the Liberals and the Communists. They all wanted increases in foreign aid levels, but as they had very diverging opinions of how to finance these increases, they did not succeed in altering the policy of the Cabinet, which was backed up by the Conservative Party.

Because of the long history of voluntary aid organized primarily by the Christian organizations, there seems to have been vague, common support in the Swedish electorate for a general policy of assisting the poorest countries. But this sentiment was in most cases neither based on a deep knowledge of the situation in the recipient countries nor of Swedish aid policy (Lindholm 1970). When asked whether they approved of the current level of aid, most respondents replied affirmatively, according to an annual survey ordered by SIDA (Styrelsen för Internationellt Utvecklingssamarbete; Swedish International Development Cooperation Agency). When asked about the details of Swedish development aid, however, few people knew the extent of the assistance given by Sweden or to what countries it was given. In fact, "... most people had not even heard of SIDA" (Andersson 1986:31).

The Liberals and the Communists were the most out-spoken proponents of increased levels of foreign aid, while the Conservative party was the most restrictive voice. The Social Democratic Party and the Center Party both had distinguishable factions advocating increased as well as decreased levels. The ambivalence within the Social Democratic Party was important, since this party dominated the Swedish political scene in the sixties. The support among the SAP voters for far-reaching increases in Swedish aid seems, however, to have been quite weak. This was completely clear in the mind of Mr. Gunnar Sträng, the Minister of Finance (Lindström 1970). Mr. Sträng was also uncertain whether aid-assisted shortcuts to development were realistic. In a conversation with the Minister for Development Cooperation, Mrs. Ulla Lindström, Sträng stated that

> The developing countries have to go through their Gustav Vasa period, just like we did: hunger, rebellion, and war. Little by little they'll cope. You'll see. Like we did... The developing countries just can't be helped. Not in the present state. (Lindström 1970)[55]

---

[55] King Gustav Vasa ruled Sweden in the 16th century.

4.5. Politicization through country selection

In addition to making assistance conditional, donor countries may also politicize their assistance through choosing recipient countries perceived as performing in accordance with donor preferences. In the first decade of Swedish governmental development assistance, recipient countries were chosen on grounds other than those with intentions to reward/punish certain behavior. As discussed above, the main criterion for receiving Swedish aid was the degree of poverty. But it was also recognized early on that Sweden – being a comparatively small country – had to concentrate its aid ambitions on a restricted number of recipient countries (Government bill 1962:100). In addition to this, continuity was emphasized. The simple fact that a particular country had previously received aid, constituted a reason for continued aid. As Sweden had been active in for example Ethiopia and Pakistan as early as in the 1950s, these countries were seen as natural recipients of the now more ambitious foreign aid flows. The fact that these countries were neither democratic, nor showed any extensive efforts to increase social equity did not change the support to these countries.[56] The selection of recipient countries had, at this time, not yet become politicized to any large extent. As the aim was so clearly focused on how to combat poverty, the policies adopted by the governments in the recipient countries were not of primary concern. The subsequent political controversies over whether aid given to a particular country was also a way of supporting its regime had not really reached the Swedish development aid agenda. They certainly did, however, in the 1970s.

4.6. Summing up on the early Swedish development aid

When summing up on the early Swedish governmental foreign aid policy, it can be concluded that it was characterized by mainly altruistic goals with a specific focus on the alleviation of poverty. In order to avoid potential domestic biases as well as trying to circumvent the disadvantage of being a relatively small donor, most Swedish aid was distributed through multilateral agencies. The potential for domestic positive effects from certain kinds of aid was recognized, but rather than advocating these, they were mainly seen as threats against an efficient allocation of the national aid resources. A possible explanation for the lack of stronger pressures from domestic business interest groups is that the few large transnational companies operating at that time were doing extremely well, and the need for state support was consequently quite small. In the latter part of the 1960s, however, this was beginning to change. The strong position of these companies on the international market was weakened, paving the way for increased pressures on the government to also take domestic "needs" into consideration.

The altruistic stance was, however, to some extent counterbalanced by the divergent interests within the Social Democratic Party (SAP). On the one hand there

[56] At this time, Ethiopia was ruled by king Haille Sellasie and since 1957 Pakistan had been governed by the military.

was a strong faction within the SAP consisting mainly of intellectuals propagating the notion of sharply increased levels of Swedish foreign aid, and on the other hand there was a group of representatives of the more traditional working class recruitment base of the SAP. The latter group's preference seems to have been more in line with those of the party's working class voters. One of the outcomes of this "lack of solidarity" (as reported in for example Lindholm 1970), was that quite large amounts were invested in efforts to increase public support for an ambitious foreign aid policy from the 1960s onwards: "No other regular budget item has used anything like these amounts to popularize its activities among the taxpayers". (Andersson 1986:34)

Before continuing the description of the historical evolution of the foreign aid policy into the 1970s, another thread will be picked up; that of Swedish national environmental policy and initiatives taken to reach agreements with other countries. As argued above, a reasonable point of departure in this area is the 1972 Stockholm conference on the Human Environment. It should be noted though, that there were some initiatives taken before this conference. Two meetings were for example held in Visby, Sweden, in 1969 and 1970 concerning oil pollution in the Baltic Sea. Furthermore, two bilateral bodies were already in existence, the Sound Committee (1960) between Sweden and Denmark and the Gulf of Bothnia Committee (1971) between Sweden and Finland (Hjorth 1992). The activities of these committees, however, consisted of specific, environmental initiatives and rather than comprehensive joint programs. Comprehensive environmental, international cooperation commenced in the Baltic Sea region in the early 1970s. The most important aspects of the cooperative efforts regarding transnational, environmental disturbances in this region took place under the auspices off the Helsinki Commission (HELCOM). This is where the focus will now be set. Other kinds of activities or processes that influenced the Swedish policy in this area, for example the domestic environmental policy, will be taken into consideration as well. Before the early HELCOM process is described, however, a short background on the contemporaneous Swedish domestic environmental policy is given.

5. SWEDISH ENVIRONMENTAL COOPERATION IN THE 1970S

5.1. The spirit of Saltsjöbaden and sectoral administration

Traditionally and in general terms, the process of policy formulation in Sweden has been described as conflict aversive. The so-called Saltsjöbad agreement marked the beginning of a period of comparably peaceful settlements between employers and labor unions, without state interventions. In the form of the *Saltsjöbadsandan* (the spirit of Saltsjöbaden) this scheme to reach mutually acceptable solutions was also applied to other areas of potential political conflict (Norborg 1988). The goal was to identify at least the most important stakeholders, and then try to find a settlement that everyone could agree upon. The treatment of environmental issues in Sweden in the 1970s does not seem to contradict this general pattern (Lundqvist 1996). The upsurge of public interest in the late 1960s and early 1970s caused by mercury pollution, visibly inferior water quality in many places, air pollution and similar issues, pushed the up-to-now quite fragmented environmental policy higher up the political agenda. The attempt to accommodate potentially conflicting interests was clearly visible. The needs of modern society ought to be balanced against what the environment "could stand". In fact, the composition of the Executive Board of the Swedish Environmental Protection Agency (Naturvårdsverket (SNV); created in 1967) was almost corporative in its nature. It consisted of identified stakeholders, such as the Swedish Industry Federation, Union interests, the Swedish Association of Local Authorities, and a few more. There was, however, no representative from any "environmental organization" (Lundqvist 1996).

Somewhat in contrast with the situation today, the environmental policy of the early 1970s was organized on a *sectoral* basis (Lundqvist 1996). Assessments of environmental impacts from different kinds of activities were not to be made by a centralized administrative body. Neither was the rise of environmental issues on the agenda perceived as implying a need for a separate ministry within the government. Rather, the effects on the environment were to be taken into due consideration where they appeared, and by the actor causing the environmental disturbance. This approach was seen as facilitating the reconciliation of different demands on the environment and natural resources. The issue was more to find a reasonable balance between competing interests, than to assess the environmental impact per se. The policy-making on these issues was made in the Ministry of Agriculture and in the Cabinet.

The picture is not clear-cut, however. The creation of the SNV in 1967, a central agency dealing exclusively with environmental issues, was a sign of a movement towards a more centralized approach. This mixed tendency towards both sectoral decision-making and centralization of environmental policy-making was not an exclusive feature of Swedish policy adaptation. Rather, the augmented interest shown in environmental issues was making additional institutional resources increasingly important.[57] Competence and resources were to be concentrated in a key, administrative body. Nevertheless, in comparison with the contemporary higher degree of centralization, the early Swedish environmental policy in very general terms could be characterized as built around a sectoral nexus. This particular feature of the Swedish policy probably complicated the work on the international scene, as it was not always clear where the competence was located and what institutions to involve. There was no natural administrative or political body, which could take the lead in the interaction with other countries in matters of transnational environmental cooperation.

Another difference from contemporary Swedish environmental policy was the relatively infrequent use of economic incentives. Typically, when lower levels of pollution or other changes in behavior were desired, administrative regulations were used to achieve this, rather than attempts to induce changes in behavior through al-

---

[57] See for example Haas (1990:133 pp) for a description of a similar development in the Mediterranean countries.

tered economic incentives. It was stated in the 1969 Environment Protection Act that activities that could cause environmentally harmful externalities required a permit. In this permit, the circumstances under which the activity was allowed to take place, such as for example maximum levels of emissions and allocation, were stated. When deciding whether to grant the permit or not, the potential societal value of the activity was compared with any harmful effects on the environment as well as on the interests of other actors, in line with the balancing of interests approach mentioned above.

5.2. The Stockholm Conference

Preceded by the OECD Environmental Committee and the international group working with transnational air pollution issues under the auspices of the UN Economic Commission for Europe (UN/ECE), the 1972 Stockholm Conference on the Human Environment was a significant indication of the increased interest in international, environmental issues.[58] At the institutional level, the most important immediate outcome of the Stockholm conference was the creation of the United Nations Environment Programme (UNEP). UNEP was set up with the task of coordinating international, environmental cooperation, as well as enhancing collaboration among Non Governmental Organizations (NGOs) across national borders. As a UN agency, UNEP also embraced member countries outside the Western sphere. Its perception of the most critical needs when it came to environmental issues preceded later characteristics among international organizations as well as government bodies. Rather than concentrating only on projects with assumed environmental implications, the goal was to attach environmental considerations to projects in all areas. In a way this was a moral imperative. UNEP should be, so to speak, the environmental conscience of the UN family.

The full sovereignty of all nation-states was recognized as a fundamental principle for international cooperation at the Stockholm Conference. The environmental implication was that each country had the responsibility for its own domestic activities and should ensure that these activities did not harm other countries or common natural resources (such as the open sea). The nation-state had the responsibility, and could not be over-ruled by any other judicial body.

During the Stockholm conference, a Finnish initiative was presented proposing the creation of a regional, environmental organization with the aim of enhancing the increasingly deteriorating environmental status of the Baltic Sea. In fact, prior to the Stockholm Conference the Finnish government had already made an informal proposition concerning environmental cooperation between the states in the Baltic Sea region, but this initiative did not result in any formal agreement at that time. Finland was in many ways deeply affected by policies adopted and actions taken in the neighboring country of the Soviet Union, and therefore had strong incentives to

[58] The UN/ECE had been established in 1947, with the primary mission of supporting the rebuilding of postwar Europe.

initialize regional, environmental cooperation. Although Finnish national security issues certainly were of primary importance because of the country's exposed geographical location, transboundary environmental problems were also high on the agenda. Due to its long border with the Soviet Union, Finland is vulnerable to all kinds of potentially environmentally disturbing activities taking place in the northwestern part of the Soviet Union. In the 1960s, a network of scientific-technical programs evolved between the two countries. These networks were not very productive, however, in terms of abatement initiatives. The primary outcome was instead scientific data compilations and the establishment of personal relations between scientists (Henttonen 1996:39). Consequently, from an environmental abatement point of view there was a need for the creation of a more action-oriented regional body with the aim of advancing agreements on pollution restrictions involving all countries adjacent to the Baltic Sea. Coinciding with this perceived need, the political incentives for the Soviet Union to go along with the initiative to create a new political forum between East and West were quite obvious (Hjorth 1992). It would give the Soviet Union a convenient opportunity to air East-West political issues. This is not the place, however, to assess the relative importance of the two converging types of incentives facilitating the institution of a Baltic Sea regional organization. It is only concluded that particularly Finland and the Soviet Union had incentives to increase the level of transnational, environmental cooperation.[59]

After two preparatory meetings in 1973, a draft convention was presented by Finland to the seven coastal states of the Baltic Sea. The draft was generally accepted, except for the issue of oil pollution from vessels, which was to be worked out during a later meeting in Sweden (Hjorth 1992:164). In February 1974, the convention was signed by the delegates, and after a Swedish proposal, an interim organization was set up to continue the work.[60]

There are two features of the 1974 convention that are of particular interest in the present context. The first concerns the Swedish perception of environmental problems in the Baltic Sea area and the second the attitude the Swedish government showed towards international cooperation on environmental issues in general.

5.3. Swedish perception of the environmental problem

The official approach to environmental disturbances in the early 1970s often focused on health hazards on the one hand, and technical solutions to the problems of point

---

[59] The International Council for the Exploration of the Sea (ICES) to some extent also influenced the evolution of environmental cooperation between the littoral states. ICES is an intergovernmental organization created for the support and coordination of maritime research. In 1970, a working group on Pollution of the Baltic was created within the ICES, which provided some of the background material used in the process of creating an environmental organization in the Baltic Sea region (Haas 1990:232 pp.).

[60] It should be noticed that to be judicially valid, a convention has to be ratified (usually by the national parliament) by its signatories. This did not happen until 1980.

source pollution on the other.[61] Problems having direct and negative effects on human health (e.g. the labeling of the 72 conference on *Human* Environment) were seen as those most important to deal with, and it was anticipated that these problems in most cases could be solved using various technical solutions. This perception was reflected in the construction of the 1974 Helsinki Convention. A large part of the convention focused on "Vessel Source Pollution" (Helsinki Convention, Articles 7-9, and Annex IV). Although this group of environmental disturbances encompasses several types of pollution, it is nevertheless comparatively distinct. Quite extensive and precise statements were made concerning, for example, whether different kinds of substances were allowed to be discharged, or not.

In contrast to the relatively distinct group of Vessel Source Pollution, the large category "Land Based Pollution" was vaguely formulated regarding what levels and types of emissions were unacceptable. In spite of this, it seems rather evident that it was the point sources that were perceived as those of highest relevance. The imprecise formulations of what measures to undertake, were to some extent caused by the political tensions between the Western, Scandinavian states and the Soviet Union (Hjorth 1992). It was difficult to attain any more far-reaching agreements. As for the almost exclusive focus on point source pollution, this reflected the predominant priorities of the Scandinavian countries. At that time, it probably was cost-efficient to at least initially focus on the point sources. A reasonable first step was for example to purify large industrial emissions and to build waste water treatment plants. The more difficult task of curbing diffuse sources of pollution was not yet perceived as an environmental problem of the same magnitude as the point sources. This, however, would change in the decades to come.

5.4. The transboundary dimension

In the early 1970s there was a growing awareness of the need for international, environmental cooperation among Swedish policy-makers. The rapidly deteriorating environmental status of the Baltic Sea was becoming more and more obvious. It was quite clear that cooperation between the coastal states was a prerequisite for a successful abatement program. The trend towards recognition of the increased interdependence among the states was, in the case of Sweden, not only limited to issues on the environmental status of the Baltic Sea. As described above, the foreign aid policy of Sweden had been expanding rapidly in the second half of the 1960s. As a relatively small country, the vulnerability of the Swedish economy to external crisis became obvious to most people (the oil crisis in the first half of the 1970s was but one example of this). The smallness also implied strong incentives for increased activity in international fora, for example in the United Nations.

The creation of a regional body to combat negative effects on the environmental

[61] A distinction is often made between point source and non-point source (diffuse) pollution. The former concerns pollution from specific effluents (e.g. industrial and municipal waste water) while the latter is focused on more dispersed emissions (e.g. pollution from agriculture and transportation).

status of the Baltic Sea should be seen against the background of the increased international interdependence in general and the rise of environmental issues on the policy agenda in the Western world in particular. There was a strong growth in interest concerning environmental issues in general in the Swedish electorate, which facilitated the acceptance of increased efforts to combat environmental disturbances in Sweden, as well as outside the country's borders. This may be seen as a structural prerequisite for intensified action when considering environmental issues in general. What made Sweden and Finland somewhat different from the other HELCOM members, however, was that it could reasonably be argued that these countries potentially had the most to gain from enhancements in the environmental quality of the Baltic Sea. The countries in this region have divergent possibilities to extract different kinds of utilities from the Sea, depending to a large extent on the stretch of the national coastlines. Sweden and Finland together possess about three quarters of the total Baltic Sea coastline (AMBIO 1990). Sweden alone possesses approximately 60 percent.[62] The longer the coastline, the more value could, ceteris paribus, be potentially extracted from the Baltic Sea. These gains would come from tangible resources such as fish catches, real estate values (since the value of real estate close to the coastline is positively correlated with the environmental state of the surrounding water), tourist activities etc. In addition, the longer the coastline, the greater the potential for using the sea as a reservoir for unwanted substances. Furthermore, for example port facilities are generally of a greater value to countries endowed with long, coastal borders. Although the domestic need for port facilities primarily is a function of the county's economy, a longer coastline means fewer transport needs on land, which decreases transportation costs, thus making the potential port facilities' value larger, ceteris paribus. Finally, it should be mentioned that both Sweden and Finland have very valuable archipelagos.

Returning to the evolution of the Swedish development aid policy, the trend of increasing activity in the foreign policy domain, regarding development aid (in the 1960s) as well as regional, environmental cooperation (in the 1970s) has been shown. The question now is whether this trend also continued regarding foreign aid in the 1970s.

6. FOREIGN AID IN THE 1970S: TO WHOM AND IN WHAT WAY?

6.1 Country programming

The foreign aid policy of Sweden in its first decade was characterized mainly by altruistic motivations, although the outcome of the policy process was influenced by the different opinions inside the SAP on the content of the aid policy as well as by

---

[62] The share for the other states is: Finland (17%), Estonia (5%), Germany (5%), Russia (4%), Latvia (3%), Poland (3%), and Lithuania (0.6%). Source: personal communication with the Swedish Meteorological and Hydrological Institute (SMHI).

external influences on the economy.

The latter half of the 1960s was a kind of economic turning point. The economic hey-days of the fifties and early sixties were coming to an end although this may not have been clearly visible at that time. Not only were the economies of the war-torn European countries catching up on Sweden, but doubts also became more and more widespread about the quality of the Swedish Model. Important groups in society were of the opinion that maybe it was not such a panacea as previously thought. Another factor of great importance was the radicalization of large segments of the political establishment and of public opinion (Palmlund 1986). The Swedish government strongly criticized the war in Vietnam, as did various Non Governmental Organizations (NGOs). These two factors may be seen as part of the reason why the formulation of the Swedish aid policy was getting more and more fraught with conflict. Strong forces were trying to re-shape the aid policy in a more radical and recipient oriented direction, while the poor performance of the economy at the same time tended to make the perceived need for taking domestic concerns into consideration more relevant and outspoken.

In 1968, the Swedish parliament decided that the goal of giving one percent of GNP should be reached in 1974/1975. Since the percentage given in 1967 was only 0.3 percent, this meant an annual increase in the aid budget of 25 percent. The contemporaneous general opinion could be characterized as positive but not very well informed. The poorest people in the developing countries were to be helped, but neither the government nor the public knew exactly how this was going to be achieved at that time (Ljunggren 1986). The focus placed on technical assistance in broad terms and separate aid projects in the 1960s, was in the early 1970s partly replaced by the new planning technique *country programming.*[63] This new way of allocating aid flows by looking more at the general needs of the so-called *program country* and less at specific projects made it easier for the Swedish Aid Agency (SIDA) to distribute a larger volume of aid. The resources used for general import support rose from a few percent in 1970 to 40 percent of the total Swedish aid budget by the end of the decade (Ljunggren 1986:49). The practice of country programming served several purposes. Not only did it make large disbursement possible to handle for the Swedish aid bureaucracy, it also made it easier for parliament to exert influence over the use of Swedish aid, since it was a lot easier to pick countries worthy of Swedish support, than to judge the technicalities of specific projects. In addition to this, the country programming technique also reflected the increased radicalization among politicians as well as ordinary people. In most Western countries, radical movements were pressing the governments to change their policies vis-à-vis developing countries in political as well as economic terms. These pressures, in combination with the requests from the developing world, were reflected also at the international level, most notably by the notion of a New International Economic Order (NIEO). The outcome of this attempt to shape the international trade patterns

---

[63] The term Country Programming was first put forward in the so-called Jackson Report, published by UNDP.

in more equal and egalitarian terms was eventually quite meager, not to say non-existent. The process in itself, however, was interesting since it showed that there were influential groups of people in the electorate in the Western countries that were deeply concerned with the perceived unfairness of the international system.

These dual and contradictory pressures of increased fairness notions advocated by the more radical movements and the alleged need to take domestic concerns into serious consideration, resulted in a somewhat inconsistent foreign aid policy. The volume of the assistance was quite large and increasing, but at the same time the foreign aid was becoming more and more influenced by demands from domestic interest groups.

6.2. Increased conditionality

The original outline of the Swedish aid policy as formulated in the 1962 Government bill, banned most attempts to blur the distinction between foreign aid motives and domestic needs. In practice, however, Swedish aid was never completely free from biases. Projects were often selected in areas where Swedish companies had strong market positions and some commodity grants were partly chosen to fit Swedish interests. But compared to other countries and future Swedish policy, the approach of the 1960s was quite free from domestic concerns.

> This rule [not to allow domestic considerations in the aid policy] was implemented with such religious fervor that Swedish exporters complained that recipient countries could buy from any country in the world, except Sweden! (Edgren 1986:59)

Ever since the adoption of the 1962 Government bill, Sweden had officially called for foreign aid without ties, and without restrictions imposed by different kinds of conditionalities. Nevertheless, as early as in 1972 Sweden introduced formal tied-aid, affecting about 15 percent of the total aid (Edgren 1986). This was to a considerable part due to political pressures from economic interests groups and the general recession of the Swedish economy; "...this decision, against which SIDA's staff lobbied intensively, was taken in response to a domestic recession..." (Edgren 1986:59)

The change in Swedish policy was that although the recipient could select from among different items, these had to be of Swedish origin. This ascertained a competitive element among Swedish suppliers in the procurement process, while at the same time making sure that domestic business interests were the main beneficiaries (Edgren 1986).

The Swedish government of that time showed a reluctance to get too closely associated with the emerging structural adjustment conditionalities implemented by the World Bank (IBRD) and the International Monetary Fund (IMF). In their assistance policy vis-à-vis Tanzania, for instance, the Nordic countries declared that they would continue their assistance programs at the same level, even if the Tanzanian government did not reach any agreement with the IMF (Edgren 1986:56).[64] This did not, however, contradict the tendency in Swedish foreign aid policy in the 1970s towards awareness that at least some conditions had to be attached to aid programs. Compared with the more highly conditionalized assistance programs of IBRD, IMF and USAID, the Swedish country programming praxis as well as the selection of recipients was, during the 1970s, sometimes criticized for being too lax, misdirected and inefficient. The outcome of this pressure was a cautious effort to reform the aid policy. Instead of strict prerequisites for obtaining Swedish aid similar to those of the World Bank and the IMF, Sweden attempted to reach a mutual understanding in the form of joint seminars, memorandums and symposiums. In other words, although the content of the Swedish requirements on the recipient country were often quite similar to those adopted by the IMF and the World Bank, Swedish authorities preferred to use its long-established "consensus policy" rather than adopting strict conditionality requirements.

6.3. The 1970s: Increasing domestic interests

Summarizing up on the Swedish development aid trends towards the end of the 1970s, there were two broad, and to some extent contradictory, pressures building up. On the one hand there was the radicalization of public opinion in the early part of the decade and on the other hand the downturn of the economy, accompanied by demands for both efficiency reforms and greater domestic consideration. The most important thing to note is that despite the idealistic upheaval of the late 1960s and early 1970s, the increase in aid volume, external pressures, domestic economic recessions and different special interests transformed the Swedish aid policy so that it operated in a somewhat less altruistic fashion. Swedish aid was, in relation to most other countries, still quite idealistic. But the ice was broken and domestic considerations were no longer banned from the agenda. As will be shown below, this trend continued and became stronger in subsequent years. Before continuing with the description of this, however, Swedish environmental policy in the 1980s will be addressed. More specifically, it will be shown that the tendency towards a more extensive interplay with other countries characterized this issue-area as well.

7. ENVIRONMENTAL COOPERATION IN THE 1980S

7.1. Recognized complexity

The environmental concern among ordinary people as well as in the scientific com-

[64] As the Swedish government at the same time made it clear to the Tanzanian authorities that any additional funding had to meet the IMF criteria, the Tanzanian government saw this as a kind of betrayal. Of course, this depends on the perspective. Tanzania certainly would like to have as few conditions as possible attached to the assistance program, while Sweden, on the other hand, criticized the IMF conditions as being too harsh (Edgren 1986:56 pp.).

munity increased dramatically in the 1980s. This anxiety was no longer exiled to environmental interest groups and a few scientists, but concerned also an important part of public opinion. One of the most significant insights was the high degree of complexity of many kinds of environmental disturbances. This increased level of complexity was seen in two dimensions; the disturbances affected a larger number of parameters and in other ways than previously thought, and the effects were dispersed in a greater, geographical area than had been hitherto appreciated. It is the second feature that is of primary interest in this context.

The sectional approach of the earlier decade was now turning into a wider problem perception. The environmental effects from different kinds of activities were perceived as more complex than previously appreciated. The "balance-of-interests" approach was losing ground to a more integrative view. Environmental effects had to be taken into consideration irrespective of whether they occurred within a particular sector or in several. The administrative implication was twofold. On the one hand, each administrative body had to have the capacity to take environmental concerns into consideration when implementing their activities. On the other hand, due to the increased complexity, in order to reach a fuller understanding and capacity to regulate behavior, environmental research and expertise had to be more centralized. The research and monitoring capabilities of the SNV were enlarged, and an Environmental Ministry was established and made the central political body at the governmental level. By the end of the 1980s, the full implications from the approach emphasizing the complexity of contemporaneous environmental problems were spelled out explicitly; "... responsibility and care for the environment should permeate all walks of life and society." (1990/91 Environmental bill, cited in Lundqvist 1996)

7.2. Policy principles and economic incentives

Another important change in the general Swedish environmental policy of the 1980s was the emergence of environmental *policy principles.* The Polluter Pays Principle (PPP), the Precautionary Principle, the Best Available Technology (BAT), and Best Environmental Practice (BEP) concepts were the most important. These principles were important precisely because they are general in their nature. The use of such notions in particular facilitates application on a wider range of issues and actors. In fact, all the mentioned principles have tended to migrate from country to country, sometimes taking the route over international institutions.[65]

A different significant environmental trend of the 1980s was the increased confidence in economic incentives' ability to influence the behavior of companies and people. As shown above, the most common regulatory instrument of the 1960s and early 1970s was administrative regulation. Now, environmental taxes, fees and different kinds of tolls were increasingly introduced in many environmental areas. By

---

[65] As will be shown in the next chapter, they have also been incorporated in the 1992 revision of the Helsinki Convention.

making environmentally harmful externalities visible and redeemable, the actors theoretically would allocate resources efficiently, thus optimize profits or utility. Interestingly, the use of economic incentives shows that the "balancing-of-interests" approach was still relevant. But the focus now shifted from trying to balance the interests of different stakeholders, to attempts to maximize the utility derived from production, where harmful environmental side-effects were made an integrated part of the production cost.

The Polluter Pays Principle has in most cases been given a primarily moral interpretation. Those causing environmental disturbances should also bear the cost of their remedy. If the goal, however, is to create incentives for an efficient allocation of resources only, it does not matter who pays, as long as the property rights are clear and enforced. Ronald Coase showed this in 1960 in his seminal article *The problem of social cost* (Coase 1960).[66] Either the actor causing the disturbance pays (when the other stakeholders have the right to an unaffected environment) or the afflicted parties do so (when the actor has a privileged right to undertake the project in question). Either way the outcome is identical. The disturbance is attended up to the level where the marginal abatement cost equals marginal utility.[67] This aspect of the PPP is especially important in studies of international relations. It is generally not possible to enforce PPP clauses in international contexts, since there is no authoritative enforcer. The result of this may be, if efficiency is still a major goal, that the PPP is transformed into "VPP, Victims Pays Principle" (Mickwitz 1998). The ethical value of this "principle" is not very impressive, but the efficiency argument is unencumbered.

At the close of the eighties it was stated that "the local, environmental problems [in Sweden] in many cases have been eliminated" and that the "contemporary environmental problems in Sweden to a large extent are caused by emissions originating from other countries" (SOU 1990:88 p. 37) It was further claimed, that most of the these environmental problems were a "concern for Northern Europe" (SOU 1990:88). From this perspective, transnational cooperation seemed all the more relevant. The process of acidification, for example, was seen as being caused primarily by emissions originating from countries other than Sweden (especially Eastern Europe, the UK, and the Ruhr area in Germany). The joint inflow of nutrients into the Baltic Sea caused by the adjacent countries was about 2-3 times higher than the estimated critical load. Pollution from international shipping was serious and seemed to imply joint action. The effects from chlorinated and bromide organic compounds, often of uncertain origins, were not yet well known, but nevertheless worrisome.

---

[66] This is a theoretical conclusion. The point here, however, does not concern efficiency per se, but rather that the grade of efficiency may not correlate with whether the polluter or the public has the right on its side.

[67] Nota bene, under the assumption of complete and perfect information and zero transaction costs.

7.3. HELCOM and the 1980 Helsinki Convention

All contracting parties eventually ratified the 1974 Helsinki Convention in 1980. There was not an environmentally related determinant behind the ratification, but a political one. The Soviet Union had made the acknowledgment of East Germany as a sovereign state a prerequisite for the ratification of the convention. When the other contracting parties finally accepted this, the convention was ratified and entered into force.

The main organizational body of HELCOM is The Baltic Marine Environment Protection Commission (the Commission from here on).[68] The Commission meets annually and the assembled national delegations try to find joint decisions on how to improve the environmental state of the Baltic Sea. All decisions have to be unanimous, which means that far-reaching commitments are hard to reach. On the other hand, this institutional design makes free-riding more difficult and thus creates incentives to put pressure on laggard countries.

The main HELCOM instrument of influence under the 1974 convention, apart from the convention itself, was the *recommendations.* The recommendations should, however, not be seen as instruments to be used under the discretion of HELCOM, but rather as an outcome of a bargaining process between the contracting parties. The countries act in accordance with their interests. The national strategies could, however, to some extent be influenced by for example data compilations undertaken by HELCOM and its mandate to suggest agendas for joint action.

During the 1980-1990 period a total of 102 recommendations were adopted (HELCOM 1997). These recommendations may be categorized into three groups: Land-based pollution, Vessel-source pollution and Clean-up activities (Hjorth 1992). It is not certain that the number of issued recommendations fully reflected the importance attached to the different groups of environmental enhancing activities. A large number of recommendations could, for example, also indicate political tensions between the parties or a high degree of complexity of the particular environmental problem. Nevertheless, it seems reasonable to assume that the salience of the issue-area is one factor of importance, and that it correlates positively with the number of recommendations. A brief description and interpretation of the distribution of recommendations on the three groups may therefore shed some light on what types of activities HELCOM focused its attention on.

The trend in the number of adopted recommendations concerning Land-based pollution was increasing in this period. During the first half of the decade, seven items were adopted compared to 30 in the second half. Regarding Vessel-source pollution, no clear trend can be discerned, in the first period 17 recommendations were adopted and 18 in the latter period. Finally, clean-up activities showed a decreasing trend. From twelve in the first half of the decade, down to four in the

---

[68] In the spring of 2000, the organizational structure of HELCOM was redrawn. The Commission continued, however, to be the most important organizational body.

second.

Table 3. Number of adopted HELCOM recommendations in the 1980-89 period.

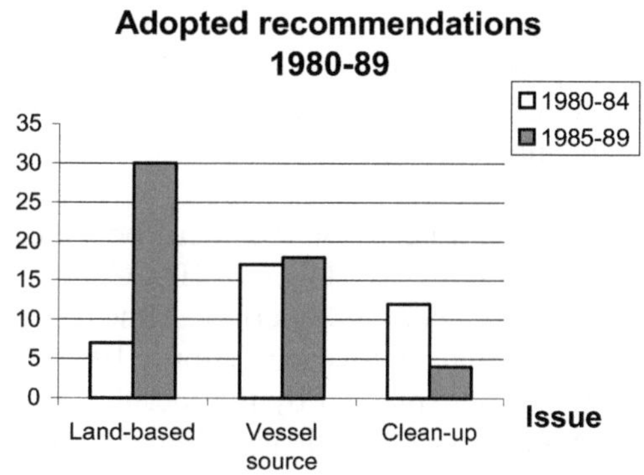

Source: Hjorth (1992).

In aggregate, it may be noticed that both the groups of Land-based and Vessel-source pollution were considerable larger than that of Clean-up activities (43, 40 and 17 percent respectively). Possible interpretations of the referred trends are:

Land-based Pollution
The increasing trend reflected a growing concern of the effect on national environmental interests from emissions outside national borders. As a large part of what was emitted close to different watercourses in the Baltic Sea drainage basin eventually ends up in the Baltic Sea, the need to also include internal waters and watercourses, which were not included in the 1974 convention, was becoming increasingly appreciated.

Vessel Source Pollution
The attempts to regulate and coordinate activities on the Seas have a long, and partially successful, history (Keohane 1984). Since no individual actor has sovereign authority, agreements on codes of conduct, monitoring and sanctioning instruments are pivotal. There are no clear trends in this group as the number of adopted recommendations was similar in both periods. Even in its first years of existence, HELCOM gave a lot of attention to Vessel-source pollution. Despite these efforts, however, the problem did not become less relevant. Rather the contrary, due to the increase in volume of traffic and the lack of clear and enforceable regulation, pollution from vessels increased.

Clean-up activities
These kinds of activities were in most cases relatively easy to define. The aim was

usually clear-cut and straightforward, to remove harmful waste from environmentally sensitive areas. Implementation, however, may be hampered by lack of domestic political and economic capacity. The interpretation in this case would be that in the first half of the decade, joint recommendations were endorsed when the more affluent Nordic countries were the ones most affected. The private incentives facing the individual actor were in these cases sufficiently strong for making implementation worthwhile irrespective of the behavior of the other parties. When the sites with the highest net returns on investments had been addressed, the propensity to adopt further recommendations diminished.

When looking at individual state's behavior, a pattern becomes visible. Regarding for example the number of state initiatives taken in HELCOM, Finland and Sweden were the most active states in the 1982-1988 period. Almost every other formal initiative was taken either by Finland or Sweden (Hjorth 1992).[69] The high level of activity by Finland may be partly explained by the fact that it was the initiator and host of the whole HELCOM-process. Nevertheless, Finland and Sweden were in this respect probably those who showed the greatest interest in the HELCOM-process.

Another indicator of the subjectively perceived preferences of the Contracting Parties that it is possible to extract from the HELCOM-process is related to the concept of *Lead Country*. The contracting parties were on a voluntary basis assigned a special responsibility for selected environmental issue-areas. It is somewhat difficult to draw decisive conclusions on the basis of "who took what", but it should be noted that Sweden during the 1980s took a particular responsibility for the issue concerning waste water treatment from urban areas. This particular area of environmental abatement required considerable investment. Swedish companies had a considerable know-how on these issues, since large domestic resources had been invested in waste water treatment infrastructure in Sweden in previous decades. As the number of municipalities in Sweden lacking waste water treatment facilities began to diminish, the industry experienced over-capacity. The large investments required particularly in the Socialist countries were certainly appreciated by Sweden, although more precise choices of governmental strategies on this issue have not been investigated.

7.4. Summing up: Increased transnational activity

Summing up, it is quite clear that Sweden in the eighties showed an increasing interest in the environmental state of the Baltic Sea. This elevated relevance also raised the propensity to engage in transnational activities such as the HELCOM-

---

[69] The total number of initiatives taken in this period was 747. Approximately 28 percent of these was taken by Finland and approximately 18 percent by Sweden. For details on this, see Hjorth (1992 Chapter 8).

process.[70] Not the least since it was becoming more and more obvious that a considerable part of what needed doing implied coordination with the other riparian states.

There was, however, an emerging perception especially among the Nordic countries, to see these activities as "too much talk and too little action" towards the end of the decade (Pedersen 1994). It was decided, that the 1988 HELCOM meeting should take place as a ministerial meeting, to emphasize the importance of improving actual implementation of adopted recommendations. At that meeting, two specific outcomes were of special importance: The 50 percent reduction declaration and the promotion of the so-called *Precautionary Principle* (PP).[71] The latter should be interpreted as a reaction towards the attitude of demanding complete knowledge, before taking action. The new PP strategy was to take precautionary measurements if a particular emission was *suspected* of being harmful. The 50 percent reduction declaration was in many ways inspired by the Second North Sea Declaration, which had been issued the year before. Even though the Baltic Sea Declaration was stated in a comparably vague manner,[72] it did in some ways set the stage for the major changes to come in the next decade. However, before these changes are described, we will return to the context of Swedish development aid, and the contemporaneous development in this issue-area.

8. DEVELOPMENT AID IN THE 1980S: GETTING IN LINE

8.1. Economic downturn

The economic performance of Sweden was quite weak in the early eighties. In common with most other Western countries and many developing countries, Sweden suffered from the effects of the second oil-shock. In addition to this, the Keynesian offensive fiscal policy of the preceding decade was coming under intense criticism. The notion of a "home-made" economic crisis gathered more and more proponents. The lesson, according to these voices, was that there were no easy solutions. The Swedish economy had to be restructured. This restructuring would affect most sectors of the economy, development aid included.

The economic setback in Sweden took place at the same time as the debt crisis in the developing countries began to surface. Many developing countries were hit quite badly when their loans from international creditors began to expire.

---

[70] The work within the HELCOM process was, however, not the only multilateral, environmental work with bearing on the situation in the region. Other relevant conventions were for example the 1979 Convention on Long Range Transboundary Pollution, the 1988 Protocol on reduction of sulphur emissions and the 1988 protocol on limitations of nitrogen emissions.

[71] The 1988 declaration by the Contracting Parties to reduce discharges by 50 percent by 1995 can be found in HELCOM (1995).

[72] Particularly the lack of a clear reference to exactly *when* the reduction should be implemented. The aim to implement the 50 percent reduction "no later than 1995" in the Declaration was only expressed as "an example". (HELCOM 1988)

It will be shown that the difficulties experienced both in Sweden and in the recipient countries resulted in important changes in Swedish aid policy. The most important change was the rapprochement to the policies adopted by the majority of the other OECD-countries, where the distinction between aid, trade, and traditional foreign policy was not always upheld.

8.2. Tied aid becomes "return flows"

As noted above, the practice of tying aid was a hot discussion topic in the 1970s. Many saw this kind of assistance to developing countries as a second rate kind of aid, although pressure from business interests made it difficult to avoid it completely. In fact, in 1972 and 1973, the percentage given as tied aid rose to eight percent, then to 14, and eventually stabilized at a level of between 15 and 20 percent. The increased use of tied aid in 1973, however, was accompanied by a parliamentary decision to use only the practice of "single tying". The procurement should be from Swedish firms but what particular provider to choose was up to the recipient country. In this way, at least some competition was upheld. The main reason for the increasing tendency to tie aid was, according to the government, the strain put by the aid budget on the balance of payments and the need for employment opportunities in Sweden (Jacoby 1986). This increased tendency of using tied aid seems to have been seen as a sad watering down of the idealism of earlier times among many of the more experienced aid workers, not least within the aid agency, SIDA (Jacoby 1986). A number of scientific studies were presented, showing the negative consequences for the recipient country from aid tying, but this did not seem to influence the policy chosen to any great extent. The structural forces pulling away from an idealistic implementation of foreign aid appeared to be stronger: "...it was a price deemed necessary to pay in order to keep business and union interests happy." (Jacoby 1986:89)

At the end of the 1970s, the concept of tied aid had become too burdensome, and the term *return flow* was therefore introduced. The notion of return flows became an immediate success, and was referred to time and again and still is today. The essence of the return flows concept is very similar to that of tied aid, with one important exception. It has considerably more positive connotations. The only real difference to the tying procedure is that the return flows concept less precisely defines how the money should return to Sweden. It is not specified what the recipient country is expected to buy, in fact, this is not even the focus of the decision. Instead, "dynamic" effects, the cooperative aspect of the interaction etc are emphasized (Jacoby 1986).

The resource flow discourse was embraced by business interests as well as by labor unions. The concept facilitated the continuation of giving Swedish suppliers a favored status in procurement of development aid goods and services. What may seem more difficult to comprehend is that its negative effects were so little debated. Apart from the positive connotations attached to the return flows concept, there were probably several explanations behind its success. The general economic hardships in Sweden raised the opinions of business and labor interests on the agenda. But there was also an increasing tendency to put forward different demands on efficiency in bureaucratic implementation procedures in a more general sense. The economic difficulties did not allow for "waste" of resources. But the efficiency primarily being focused was not that of avoiding biases in aid procurement, but rather a technical interpretation of the word, where a particular project was seen as efficiently implemented when the outcome was an easily observable, physical construction of some kind. The more difficult task of analyzing to what extent this outcome complied with the aims of Swedish development policy was given considerably less attention.

Another reason behind the acceptance of the return flows concept may have been the weakening of the radical political stands of the 1970s, in Sweden as well as in the rest of the Western world. The fast growth of the Swedish aid volume, finally, put great pressure on the implementing agency, SIDA. The institutional capacity necessary to guarantee an efficient administration of the allocated aid resources simply was not enough for engagements in general policy-formulation issues. Anders Forsse, who worked for SIDA from its early beginning and was Director-General between 1979 and 1985, formulated the lack of capacity problem in the following way:

> In Sweden... there is a shortage of administrative capacity in the several government agencies that administer development aid. This makes the powers-that-be inclined to look for forms of transfer to poor countries, which require much money, but little administration. (Forsse 1986:260)

The downside of the rather imprecise notion of return flows was the difficulty of measuring to what extent resources were actually flowing back. According to SIDA's own data, in the 1970s the return flow rate was approximately 40 percent. Thus, a considerably higher figure than what was labeled as tied aid at that time (15-20 percent). In 1990 the corresponding figure was about 50 percent.[73] This was somewhat higher than for e.g. Norway and the Netherlands (33 and 37 percent, respectively) but lower than for e.g. Germany and Finland (69 and 62 percent respectively). What is particularly interesting in this context, however, was the relation between the return flow and the level of tied aid. The percentage of tied aid was lower in Sweden than in most other countries, including those mentioned above.[74] But the quote between the return flow and the share of tied aid was particularly high in the case of Sweden (2.4, as compared with e.g. Norway (0.85) and Finland (0.86)) (Hellström 1994). This divergence could be explained in two ways. The untied part of the Swedish aid could have brought about a relatively high return flow, but it is also possible that the comparably small tied part gave rise to an exceptionally high return flow. Since the Swedish untied aid had been relatively free from conditionalities, the latter explanation seems more plausible. Or in other words, a sign of a distinction between different kinds of assistance became visible. One based primarily on altruistic grounds and one on mutual benefits.

---

[73] Figures cited from Jacoby (1986:90).

[74] The figures on tied aid for these countries were in 1990 as follows: Norway, 39, the Netherlands, 49, Germany, 66 and Finland 72 percent.

By the end of the decade, the radical political wind of the 1970s was rapidly changing into one of interdependence, free trade and liberalization. This was reflected in Swedish aid policy particularly in two respects. First, an increasing share of the total assistance was used for bilateral aid. Although this had been a trend since the mid-sixties, it became even more pronounced in the 1980s. Second, the emergence of the so-called *multibi assistance* and the diminishing gap between Swedish policy and the so-called Washington consensus were both clear reflections of a Swedish rapprochement to mainstream Western aid polices.[75]

8.3. Multilateral, bilateral and multibi assistance allocations

In the first decade of Swedish development aid, multilateral assistance was given almost exclusive priority. In 1960, almost 80 percent of the total aid budget was targeted to multilateral aid organizations (SOU 1991:48). By the end of the 1980s, the percentage of multilateral assistance had shrunk to only about 40 percent (multibi included) (SOU 1991:48 p.53). It does not necessarily follow, however, that the importance of bilateral assistance had grown at the expense of the multilateral. Although the percentage of total aid had decreased significantly, Sweden was still a relatively large contributor to the UN family in absolute terms. In some of its specialized organizations (e.g. UNICEF, UNRWA, UNDP and IDA) Sweden played a significant role, due to its comparatively large contributions. Sweden has always emphasized the importance of the UN system in official statements issued in domestic arenas as well as in international fora (SOU 1991:48). What seems to have happened was that there was a tendency to diversify Swedish assistance, in order to make it more in accordance with *both* foreign policy goals and development aid ambitions. The different UN branches were not given equal support, at least not in monetary terms. The UN was used when the policy of one of its specific units was in accordance with Swedish priorities and when it gave access to specialist competence not available within SIDA. One way to do this was to give selected UN agencies, the ones that were seen to be closest to the Swedish policies and/or perceived as more efficient, extra funding. This kind of assistance became known as *multibi assistance.* The rationale behind this practice was quite straightforward. Providing extra funding to a particular agency gave stronger influence over its policies. Although there is no direct relation in the UN between donation of funds and influence, as in for example the World Bank Group, the more one gives, the more one gets. The use of multibi assistance was in aggregate quite extensive. In 1989, about one third of the total Swedish, multilateral assistance, was directed to aid recipients in this fashion. This

---

[75] Multibi assistance refers to the practice of giving selected agencies (e.g. UNDP) within a multilateral organization (e.g. the UN) extra funding, in addition to the regular member-fee paid to the mother organization. It should be noticed that the World Bank policy referred to here does not include that of IDA. Sweden supported IDA from the very beginning, and has been an important member of it ever since, due to its relatively large contributions to its funding.
The Washington Consensus meant, very briefly, that the need for structural adjustment in recipient countries was appreciated, and that this need made conditionality in the granting of aid legitimate.

put Sweden in fourth place among providers of multibi assistance (SOU 1991:48 Ch. 18).

This kind of distribution of aid contributions was later criticized on the grounds that it renders the administration of the multilateral agency more difficult, and threatens to pull its practices away from policies jointly agreed upon (SOU 1991:48). As long as Swedish priorities did not coincide precisely with those of the UN agencies, however, the temptation to influence the latter in this way remained.

8.4. Getting closer to the Washington consensus

During the 1970s, Sweden was quite critical of the Washington consensus in general and of the policies of the World Bank and IMF in particular. The continuation of lending to Chile shortly after the coup against Allende in 1973 and the suggestion to initiate lending negotiations with South Vietnam in the early 1970s, when the war was still raging, are only two examples when the criticism by the Swedish government was very fierce.[76]

The stark Swedish reaction on the perceived lack of solidarity and neglect of human rights could partly be explained by the radicalization of Swedish public opinion in general in the 1970s. However, this change of public, political opinion was not specifically Swedish. In fact, it emanated from the United States and Continental Europe, and only later affected Sweden. What seems to have been the difference between Sweden and most other OECD-countries, however, was the urge to keep the traditional support to the poorest LDCs free from commercial considerations. Not that the benefit from international trade was denied, on the contrary, Sweden was an out-spoken proponent of trade liberalization, but also of the notion that business and aid policies should be kept as two distinct entities, requiring different considerations.

In the early 1980s, the official emphasis on keeping the two subjects clearly apart began to weaken. The pressures put on the aid budget from the economic slump began to modify the official standpoint:

> ...industrialized countries have a considerable non-utilized production capacity and high unemployment... In this situation aid assumes a special dimension as an instrument for global counter cyclical policy, the purpose of which is to speed up the world economy." (Budget bill, 1980/81:22-23)

The door was slightly opened for de-emphasizing the former clear distinction between aid, trade and foreign policy. In the budget bill of the following year, the distinction was further blurred:

> The basic motive for aid is moral. But even the enlightened self-interest of industrialized countries should lead to the same conclusion, namely ... increased resource transfers... [budget bill 1981/82:25]

As the distinction between trade and aid was increasingly blurred and the focus was

[76] Sweden had an on-going assistance program with the Allende regime, but ended it due to the coup.

put more on technical efficiency and the need for structural adjustment policies in the recipient countries, the criticism against the World Bank faded. In the budget proposal of 1983/84, the term "structural adjustment program" was explicitly referred to for the first time. The need for some conditionality was appreciated, although a cautionary attitude was adopted. The effects of the programs on vulnerable groups were to be closely monitored while implementing the project. By the end of the decade, the Nordic stand taken on the issue of structural adjustment and conditionality in granting assistance was very close to that of the World Bank and IMF.[77] Sweden, as a bilateral donor, could hardly administer any structural adjustment plans, but conditionality in the giving of aid was definitely on the rise. The convergence between Swedish and World Bank policies was, in Sweden, seen as a confirmation of the accuracy of Swedish policy:

> It is worth noting, that the Bank policy on the criticized issues actually has changed … in a – from a Nordic point of view – proposed direction (SOU 1991:48 p. 93).

It is true that the World Bank had grown sensitive to the critique of neglecting for example negative effects on vulnerable groups of people and the environment during the 1980s and to some extent changed its policies in accordance with this critique. These changes, however, were most likely not due to the Swedish criticism. Rather, there had been a drift towards appreciating these needs in most OECD-countries, including Sweden, most notably with respect to the environmental impacts of certain development projects. Being a small country, Sweden may have been more influential on these issues than expected, but it hardly had any major influence over the formulation of the new World Bank policy. An alternative interpretation could be that although the approach towards development issues of the World Bank had "softened" in certain aspects, the main change was that of *Sweden* towards the policies of the Bretton Woods institutions, as well as towards most other OECD-countries. There were still some differences, but these were relatively minor compared to the situation of the 1960s and the 1970s. The formulation of comprehensive reports on what was needed for development to take place, for example, were generally no longer seen as interference in the domestic policies of the recipient countries (SOU 1991). Aid programs conditioned by political changes in recipient countries were, by and large, seen as a prerequisite for efficient development aid. In fact,

> Today, there are no principally important differences between the Swedish and the World Bank perception when it comes to the economic and development polices of developing countries. (SOU 1991:48 p. 103)

8.5. Dichotomization of Swedish aid

Finally, we will return to the outcome of the attempts in the 1970s to change the

---

[77] Sweden does not have any representation in the Bretton Woods Institutions IMF and the World Bank. Instead, Sweden, together with the other Nordic countries, has a joint delegation there. In the early 1990s, the Baltic States got a similar representation.

structure of the interaction between the North and the South into a New International Economic Order (NIEO). It will be shown that the pressure from business interests, labor unions, certain political groups and the tight governmental budget of the early 1980s did not *only* bring Swedish development aid more in line with the mainstream OECD-countries' approach. It also resulted in a dichotomization of Swedish aid into two different categories. One of these was "traditional", thus focusing more narrowly on the poorest and most vulnerable groups. Another admitted somewhat less needy aid recipients and also took into consideration the impact on Swedish foreign policy interests and domestic priorities into consideration.

Entering the 1980s the NIEO was, if not forgotten, at least considered naive and impossible to implement.

> There will be no New International Economic Order, and no real massive transfers of resources on concessional terms from developed nations to poor, underdeveloped ones… I think it is important for the latter to cease their hopes on any such thing, and for the former to cease pretending that they believe that any such event will occur. (Forsse 1986:250)

Although the NIEO initiative did not leave any lasting imprints on the levels of resource transfer, it did affect the transformation of the Swedish aid policy into less restrictive directions, allowing also for mutually beneficial projects to take place within Swedish aid. A Parliamentary Commission on Development Cooperation analyzed the implications of a NIEO policy shift in 1977 (Jacoby 1986). The outcome from this commission was, in line with its instructions, a proposal for a more general policy vis-à-vis *all* developing countries (a so-called *LDC-Policy).* Even though it was stated in the report of the commission that all aid resources should be spent on solidarity with developing countries, incorporating everything that could have bearing on NIEO issues into a general policy tended to obscure the distinction between altruism and mutual gains approaches.

The formulation of guidelines for the Swedish policy towards developing countries resulted in a widening of what should constitute Swedish aid. However, this did not primarily affect the traditional aid recipients. In these countries similar needs were also addressed after the policy reformulation. But, a more "advanced" form of cooperation was legitimized, which could only be implemented together with somewhat more developed countries. In other words, also countries other than those belonging to the LDC-category had become eligible recipients of Swedish aid resources:

> First of all, the rather vague outlines of a general Swedish LDC-policy were concretized into the somewhat firmer concept of *broader cooperation.* A number of developing countries – also outside the group of major aid (program) countries – belonging mainly to the lower middle-income category were to benefit from cooperation with Sweden in a wide range of areas and in forms different from those prevailing in the traditional aid relationship. (Jacoby 1986:93)

It may seem that the drift in policy terms from aid towards emphasizing cooperative aspects implied an increased need of also taking other parties' interests into consideration; a need for mutual concessions and favors. In this case, however, quite the opposite was the result of the policy shift. The emphasis on cooperation made it both possible and more legitimate to make Swedish participation more highly conditioned on the behavior of the other part. Since cooperation, unlike altruistic aid, was based on expectations of mutual benefits, canceling funding of a project, for example, is perfectly legitimate, if one expects few net benefits from it. Additionally, widening the criteria for being potentially relevant as a recipient of Swedish aid made the number of possible recipients considerably larger, thus making a selection based on Swedish priorities both necessary and more justifiable. The recipients of Swedish aid have always been selected on certain criteria, but a larger group to choose from made discretion easier.

In fact, Sweden had also earlier provided aid to lower middle-income level countries on political grounds. A few examples are the assistance to the Allende-regime in Chile, the anti-apartheid regimes surrounding South Africa, Cuba in the 1970s, and most notably the support of the Sandinista National Liberation Front (FSLN) regime in Nicaragua in the 1980s. These support programs were possibly in line with the goals of Swedish aid, although this was a highly debated issue at that time, but the point here is that Swedish aid was used for foreign policy motives rather than for poverty alleviation. These motives may, or may not, have been based on altruism. In some cases, these programs could be interpreted as attempts to influence recipient countries' policies, rather then aiming at alleviating poverty or facilitating economic development. In other cases, the support emanated from other rationales, but its content was different from what had been customary before (i.e. transfer of technology and know-how, pre-feasibility studies, joint ventures and cultural exchange programs). The amount transferred in this kind of cooperation was low in the beginning, approximately 0.3 percent of the total Swedish aid in 1979/80, but increased rapidly during the eighties (Jacoby 1986). Only six years later the percentage had risen to almost thirteen. This drift of foreign aid policy was an important background to the introduction of Swedish aid to the Baltic States in the 1990s.

8.6. Administrative changes

In order to implement these quite extensive cooperative engagements, it was necessary to reform the foreign aid administration. Due to the pressure from business interests and the changing policy winds, SWEDFUND (Swedish Fund for Industrial Cooperation with Developing Countries) was established in 1978, with the mission to make joint ventures between Sweden and developing countries easier to initiate.[78] The idea behind the creation of this fund was to increase investments in joint ventures with companies in developing countries, as previously such investments had been relatively modest. Approximately 23 million US dollars were at the disposal of

---

[78] SWEDFUND, together with IMPOD (the Import Promotion Office for Products from Developing countries) and the Industry Division of SIDA, was merged in 1991 into SWEDCORP – Swedish International Enterprise Development Corporation. This new agency had a somewhat different purpose than SWEDFUND, see Chapter six.

SWEDFUND as a base funding. As the purpose was to raise additional funding in the capital market, the Fund could still support about 30 industrial projects in its first decade of existence.

Another administrative body set up under similar auspices was BITS (Swedish Agency for International Technical and Economic Cooperation). Although it had existed on an experimental basis since 1975, it was formally established in 1979 (Jacoby 1986:94). From a relatively moderate start, the activity of BITS increased steadily. Engaging in activities such as technical cooperation (e.g. training, technical assistance, personal exchange, pilot studies) and financial assistance (concessionary credits) it had in the mid-eighties reached a turnover of approximately 60 million US dollars. As will be shown below, BITS would half a decade later become one of the most important agencies managing environmental support to the Baltic States.

The institutionalization of SWEDFUND and BITS in the late 1970s was the first apparent sign of a split of Swedish aid into two different branches. One dealing with traditional aid programs, directed mainly to basic needs in LDCs, and one dealing with more "advanced" kinds of cooperation with countries slightly better off. In a way, this was quite a natural development, since it satisfied the most important actors with a stake in Swedish aid resources. Business and labor interests welcomed it, as it increased the volume of aid disbursements in Sweden. It also pleased those who had been criticizing the aid policy and administration for being inefficient. Furthermore, it also pleased NGOs and the large groups of ordinary people who still thought that Swedish aid ought to benefit the poorest people in the poorest countries. Finally, not the least important was that this dualism in the aid policy also satisfied the SIDA bureaucracy. SIDA was at this time an outspoken critic of the new direction Swedish aid was taking. The general opinion in the agency seems to have been that Swedish aid should be kept "pure" and was not to be diluted into projects aiming at goals other than the traditional ones.

In a not too uncommon way in Swedish political history, the tendency had been to accommodate different interests rather than to let them compete openly. It should be emphasized, however, that the tension between altruistic considerations and moral convictions on the one hand, and domestic priorities and "sensible needs" on the other had not completely evaporated. The interests pulling in different directions were close to the surface during the rest of the decade, where the proponents of the various policy recommendations were always quite anxious about losing ground to one another.

> The persistent dichotomy between commercial considerations and developmental objectives surfaced sharply and has to this day remained one of the most difficult to bridge. (Jacoby 1986:95)

9. CONCLUSION

Swedish development aid changed quite significantly during the three decades 1960-1990. In the beginning its approach was rather idealistic compared to later aid policies. Time and again it was emphasized in the most important foreign aid document of that decade, the 1962 Government bill, that Swedish aid should be "pure". No biases in terms of bringing domestic considerations into the design of aid programs should be allowed. Also in practice, the Swedish profile was rather altruistic. This was primarily due to the prevalent views of the key policy-makers and administrators, the still comparably modest size of the commitment, and the prosperity of the economy.

In the early 1970s, the general political mood had become more radical, but now this was counterbalanced by another tendency, which stood in conflict with the ideal of "pure" aid. The economic outlook was now less prosperous due to external shocks such as the oil crisis as well as from domestic, structural deficiencies. Together with the sharp increase in the aid volume, defenders of pure aid were pressured by industry and union interests. The repercussions of the aid programs on the domestic economy were no longer easily dismissed. Motives other than pure altruism began to influence the formulation of Swedish development aid policy.

Paralleling this development in the foreign aid arena, increased attention was paid to environmental issues. The creation of a central, administrative agency, SNV, in 1967 was a sign of the augmented importance attached to environmental issues in general. But it was also a reflection of the common way of defining policy in Sweden, letting a central authority balance potentially conflicting interests. With the advent of the Stockholm Conference on Human Environment, the international perspective was given particular attention. Of specific interest in this context was the Finnish Helsinki Convention initiative. With the agreement on the Helsinki Convention and the institutionalization of an *Interim Commission*, a clear focus was placed on the regional need to enhance the poor environmental state of the Baltic Sea. Swedish environmental policy was no longer confined to national borders. The realization that the environmental state of the Baltic Sea was an issue that could not be solved by Sweden alone provided the necessary impetus for the emergence of transnational, environmental cooperation initiatives. The HELCOM process was a reflection of this.

Entering the 1980s, Swedish development aid policy became increasingly affected by other interests and considerations than those governing earlier assistance policies. The traditional idealistic and altruistic stance came more and more into conflict with the perceived need to accommodate less philanthropic interests, as expressed by business and labor organizations as well as by people worried about "inefficiencies". This conflict was to some extent solved by letting the aid policy evolve in two different directions. One keeping the traditional focus on combating poverty and educational needs in LDCs, and one emphasizing cooperation and mutual benefits. The first branch satisfied the traditionalists and the second pleased competing stakeholders. The institutional reflection of this split was the creation of SWEDFUND and BITS in the late 1970s, whose field of operation was activities aimed at obtaining mutual benefits. The changes in environmental policy could be characterized as a movement towards a more complex problem perception, both vertically (i.e. affecting more parameters than previously understood) and horizontally (geographical scope). One reflection of this tendency was the increased interest in the environmental state of the Baltic Sea. The insight that the eco-system of the Sea was more deeply affected than hitherto appreciated and an increased readiness to extend the perspective beyond Swedish borders made the relevance of HELCOM and the Helsinki Convention considerable. All contracting parties had ratified the convention in 1980, but its efficiency was severely hampered by the cold-war tension between the Western states and the socialist bloc. The potential for a more active Swedish regional Baltic Sea policy had become greater. In many ways, the stage was set for the distinct increase in ambition levels, which materialized in the 1990s. The situation was similar in the other more affluent countries in the region (Denmark, Finland and Germany). In fact, entering the 1990s, the environmental and foreign aid policies had, in certain aspects, evolved into somewhat of a common interface.

Since the early days of Swedish development aid, its implementation had been guided by four goals: to enhance (*i*) economic growth (*ii*) socio-economic equality, (*iii*) economic and political independence, and (*iv*) democratic development (Utrikesdepartementet 1994). In 1977, however, a fifth goal concerning the sustainable use of natural resources and protection of the environment was proposed to the parliament.[79] This bill was rejected, however, as the government view was that environmental considerations should be incorporated into all sectors instead of being separated as another assistance goal (Elgström 1997). The issue was raised on several subsequent occasions, but it was not until 1987 that the environmental goal was accepted by the parliament, and added to the four others. At this time no conflict was perceived between introducing the fifth goal and integrating environmental concern in all areas of official assistance. Rather, the addition of an environmental target should be seen as a reflection of the increased importance given to environmental protection, an insight into the potentially harmful effects on the environment from "traditional" aid projects, and "globalization" of environmental disturbances. External influences probably also had considerable effects on the Swedish policy. The World Bank Group had also begun to consider environmental issues related to project implementation and the final report of the Brundtland Commission *Our common future* got a lot of attention, just to mention two important sources of influence (SOU 1994:102).

In the following chapter, the formulation of the Swedish environmental support to the Baltic States in the 1990s will be reviewed and put in perspective. The development in both the foreign aid and environmental cooperation sectors had made it possible to initiate a rather larger assistance program based on mutual gains. It will be shown that there was almost a complete consensus among the political parties in

---

[79] The first proposal came from the universities of Gothenburg and Uppsala in their comments on the Government bill 1977/78:135. Soon after, two parliament members made a similar suggestion from the Center party in the bill 1977/78:1911. See also Elgström (1997).

the Swedish parliament on both the general need for a CEE aid program and on its content.

CHAPTER 4

THE SWEDISH ENVIRONMENTAL ASSISTANCE CONSENSUS

1. THE IMPORTANCE OF HOMOGENEITY OF PREFERENCES

There has been a rather extraordinary political consensus in Sweden around the existence and design of the environmental support to the Baltic States. To show the high degree of concordance, the government policy in this issue-area will be described, as well as how the other political parties of the parliament have interpreted this policy. As the cabinets changed twice during the 1991-1996 period, all political parties in the parliament except the Left party and New Democracy have been both in government and in opposition during the period under study.[80] The description of government policy is mainly based on official documents.[81] This does not mean, however, that influences from other actors are neglected, as these are reflected in the political standpoints of the political parties to a considerable extent. It is assumed that the most important interests of influential actors in one way or another have been internalized into the agendas of at least one of the political parties in the parliament.

The ambition in this chapter primarily is to show the large degree of consent regarding both the existence of the support per se and the general features of the Swedish assistance. This is of vital importance in the application of the model on the data material, since the larger the degree of homogeneity in domestic preferences, the more reasonable the treatment of Sweden as a unitary actor is. If it cannot be shown that there has been a high degree of consent among key decisions-makers, the assumption of Sweden as a unitary actor is problematic.[82] Conflicting interests might produce intransitive, social preferences and possibly irrational governmental behavior, despite the assumption of individual rationality. It would then be necessary to analyze the interaction of the domestic interests first, and then use the outcome of this game as input in the game at the international level. In other words, a two-level

[80] The Cabinet of the Social Democratic Party (SAP) was replaced in 1991 by a non-socialist coalition consisting of the Conservatives (Moderaterna), the Liberals (Folkpartiet), the Centre Party (Centern), and the Christian Democrats (KDS). In 1994, the SAP returned to the Cabinet.

[81] When quotations from official documents are used, the author has made the translation from Swedish to English.

[82] The approach chosen here bears some resemblance to what has been suggested in Niou et. al (1989). However, while these authors try to find general requirements for making it possible to treat collectives, such as for example states, as unitary actors, the task here is to provide arguments in favor of the validity of this approach in this particular case study.

games approach would be required.

The assumption of the state as a unitary actor has been very common in the studies of international relations. Some of the more prominent writers are Morgenthau (1973), Waltz (1979), Gilpin (1981), and Zagare (1987). As noted by Bueno de Mesquita and D. Lalman, however, "What precisely is meant by a unitary actor is rarely spelled out in much detail by researchers who find this assumption convenient." (Bueno de Mesquita & Lalman 1992:27) That is an important reason why considerable space in this study is devoted to the reasonableness of the unitary actor assumption in the context of Swedish environmental assistance to the Baltic States.

It might be suggested that instead of relying on homogeneity of preferences, one could fall back on the argument that evolutionary processes in the long run will assure adaptation to the environment (i.e. rationality) also without assuming any intentional decision-making. When it comes to social groups such as the nation-state, however, this is a very dubious suggestion. According to Robert Nozick:

> A conservative presumption in favor of institutions, traditions, and biases that have existed for a long time is extremely implausible on its face, unless it is severely restricted. Slavery existed for a long time, and so has – these continue – the subordination of women in society, racial intolerance, child abuse, incest, warfare, and the Sicilian Mafia (Nozick 1995:130).

This statement points to the main reason why it cannot be assumed that composite units such as nation-states are rational on evolutionary grounds. The nation-state is not adapted to its environment in any general sense, but rather to the strongest interests of the dominating sub-groups *within* the state. The analogy between markets and political systems is in most cases simply not valid. Firms in a competitive market are subject to a rather efficient selection (price) mechanism, where "irrational" corporations will not survive. There is no such efficient mechanism shaping the evolution of political institutions. If a political institution is not adapted to its surrounding society, there is no direct mechanism to make it wither away. Rather, it will continue its existence as long as is has strong enough supporters. It is true that vote-maximizing political representatives in a democracy do have to adapt their policies to the majority of the electorate, but this link between individual voters and the survival of political institutions is nevertheless a lot weaker than, for example, the structural price mechanism.

As discussed in Chapter one, if one draws the inference that the rationality assumption at the individual level implies that the collective (e.g. Sweden) can also be assumed to act rationally, this may cause an ecological fallacy. A statement that is true at one level might not be so at another level. It should be recapitulated from Chapter one, that Kenneth Arrow showed in the 1960s that there is no reasonable and democratic procedure to aggregate preferences in a democratic way in such a way that a collective transitive preference order is guaranteed.[83] And if we do not have a transitive preference order, the assumption of rational behavior is not valid at the collective level. But, if it can be shown that the preferences of the group are ho-

---

[83] Arrow showed this in the so-called *Impossibility theorem* (Arrow 1951).

mogenous, this problem disappears, since the preferences of the individual and the collective will be identical. Although a perfect homogeneity cannot be shown regarding the Swedish environmental assistance, a high degree of consent among the most important domestic actors will be enough to make it defendable to treat Sweden as a unitary actor. The existence of this high degree of consent will be shown in this chapter.

The presentation in this chapter will also serve as a means to show the relevance of self-interest in the strategy of Swedish environmental assistance. Although the material presented here is of a qualitative, empirical character, this by itself does not imply that a lower credence should be attached to it when judging the over-all validity of the analysis. Rather, the objective is to complement the quantitative data in Chapter seven with empirical material revealing the formulation of preferences in documents, statements, and similar types of sources, rather than what actually has been implemented in terms of real investments. It should be noticed, however, that the use of similar data for different purposes (e.g. qualitative and quantitative analysis), might be problematic. An academic discussion has been going on for quite some time now, concerning the so-called *issue of revealed preferences*. Although this is not the place to get too deeply involved in a rather technical debate, a few remarks need to be made on this subject in order to justify the choice of also including more qualitative parts in this study. The background is that in 1938, the influential economist P.A. Samuelson invented the concept of revealed preferences. This approach has been criticized by many scholars and downright dismissed by others.[84] However, an easy dismissal of the concept is dubious, and may be based on a misunderstanding. F. W. Scharpf writes in his book *Games real actors play*:

> ...whatever may be its status in economic theory, if used as a methodological precept in empirical policy research, it could produce only tautologies instead of explanations, since observed actions are explained by preferences that in turn are inferred from observed action (Scharpf 1997:60 and note 8 at p. 67).

The critique of the concept of revealed preferences is directed at its alleged tautological features. When discussing the rational choice approach, this does not appear to be correct. It is unclear, how something that may be a logical deduction in one context (i.e. neo-classical economic theory) could be anything else in another context (i.e. empirical policy research). Logical truths can only be based on their assumptions, not the context where they are applied. When dealing with strictly logical and unequivocal models such as those of rational choice, the laws of logic *guarantee* the correctness of the model, *provided* that the assumptions are correct (Tsebelis 1996). This may seem a remarkable statement, and in a way it is. It says that provided that the laws of logic are accepted as valid, we need "only" to test our assumptions, since no deductions can give any new knowledge that was not inherent in the assumptions. Modeling then becomes a way to analyze the assumptions and formulate testable hypotheses. This means, that using similar data both for the con-

---

[84] For interesting examples of this critique, see Sen (1977) and Opp (1999). For a more straightforward dismissal, see for example Scharpf (1997:60). A review of this issue is available in Lewin (1996).

struction of a utility function and for model testing is not tautological, since the same things is, in fact, tested, namely the assumed utility function.[85] In this case, this would mean self-interested preferences. If exactly the same dimension of a single data material was used twice, the procedure is not quite tautological, but rather an unnecessary double-test. On the other hand, if two different dimensions of a data collection both representing the same process are used, one of a more qualitative nature and one of a more quantitative, it would be possible to present a stronger case.

In this study, two different dimensions of similar data will be used. One describing the Swedish assistance in a contextual way, including factors that might have affected the policy in a more general sense, and one more quantitative dimension where the focus is put more strictly on the actual project selection and design. In this chapter, the former locus is adhered to.

2. POLITICAL CHANGES IN THE BALTIC STATES

The possibilities for making far-reaching environmental commitments which included all the countries adjacent to the Baltic Sea were quite infinitesimal, as long as the Soviet Union was intact. In the autumn of 1989, this changed. The Berlin wall was torn down, and the political structures of the Central- and Eastern European countries were altered in their most basic fundamentals. In the Baltic States, as well as in most of the CEE countries, the diminishing political capability of Moscow was paralleled and reinforced by popular movements in different domains. In the Baltic States, the environmental domain became one of the most important arenas for showing dissatisfaction with the Moscow's rule. The environmental deterioration was not only important by itself, but was also perceived as a reflection of the oppression exerted by Moscow, and served as a catalyst towards greater sovereignty in the Baltic States. Previously, anti-Moscow sentiments primarily had been expressed in the cultural domain. Now, environmental concern came into focus. Protests against the hydroelectric plant in Plavinas, Latvia, surfaced in 1986 and soon became widespread. At about the same time, similar widespread popular opinions reached the public domain in Estonia, concerning the planned excavation of phosphorite in the northern part of the country. Both these large, centrally directed projects were subsequently canceled, due to the popular resentments they engendered (Taagepera 1993). In addition to the perceived environmental gain from the nullification of these projects, it was shown that popular indignation actually could affect decision-making in Moscow:

> The mass mobilization effort fueled confidence in the possibility of successfully opposing the designs of the bureaucratic regime. That in turn heartened the national demonstrations that began in 1987. (Taagepera 1993:305)

The synergetic effects from, on the one hand, the relaxed resistance from Moscow

---

[85] Instrumental rationality is part of the basic assumptions in the model.

due to the glasnost and perestroika policies, and on the other hand the increase in popular demand for greater independence, changed the Baltic political scene rapidly (Smith 1994). At first, the popular fronts were aiming at greater sovereignty but it was soon realized that complete independence was attainable (Lieven 1993). In the spring of 1990, a de facto declaration of independence was issued by Lithuania, and somewhat later, Estonia and Latvia followed suit.

The stage was now set for a more active Swedish foreign policy towards the former socialist states across the Baltic Sea. The potential for more intense interaction was not restricted to the region adjacent to the Baltic Sea, but quite soon this was where the Swedish focus was set. Numerous attempts were made to coordinate individual strategies to promote cooperation with the countries in transition. In this early phase, however, an intuitive and sometimes somewhat arbitrary approach was adopted, where factors such as proximity, cultural ties, political similarities, and the domestic, political situation were the main determinants.

3. THE FIRST THREE-YEAR ASSISTANCE PROGRAM

In the autumn of 1989, a new, more active strategy towards the rapidly transforming former socialist countries began to materialize. In the government budget bill of 1989/90 a sharp focus was placed on the political turmoil in Central and Eastern Europe (CEE). The increased potential for a more active foreign policy due to the decreasing East-West tension was appreciated. The implications on Swedish security issues as well as on trade and cooperation were at this point quite unclear, however. Hence, there were few remarks in the bill describing the details of the policy change. Nevertheless, the immediate stance taken was to increase cooperative activities: "It is the ambition of the government to further a broad European cooperation and to extend the exchange between East and West." (Government bill 1989/90:100 p. 146) It was already assumed at this stage that most of the support should benefit the countries closest to Sweden: "Sweden appreciates that the cooperation with its neighbors around the Baltic Sea – not the least with the Baltic States – is getting broader, as well as deeper." (Government bill 1989/90:100 p. 10)

A three-year program for the assistance to the Central and Eastern European countries was drafted in the Swedish Cabinet Office in late 1989, and the proposal was accepted by the parliament in the spring of 1990. A total of 1 billion SEK, approximately 2.5 percent of the total aid budget, was set aside during a three year period, 1990/91-1992/93, to "further the process of democratization and reform" and to promote cooperation between Sweden and the former socialist countries at the rim of the Baltic Sea (Government bill 1989/90:100 p. 65). Nine hundred millions were drawn from the regular aid budget, while the rest was acquired from other sources.

The environmental impact aspect of foreign aid had achieved quite a prominent place in the Swedish policy discourse by the turn of the decade. The importance of integrating environmental concern in all kinds of assistance was discussed at length in the budget bill of 1989/90. Somewhat surprisingly though, environmental considerations were not addressed in connection with the new CEE program. When the new environmental foreign aid goal from 1988 was referred to, it was only done so while discussing the support directed to developing countries and not to the countries in transition. The importance of the 1987 Brundtland report and its implications on, for example, the need for incorporating environmental concern into the instructions of international institutions, was especially emphasized in the bill. There were, however, no referrals to the proposed support program to Central and Eastern Europe. When discussing the planned 1992 UN conference in Rio, it was noted that the focus there should be firmly put on "global environmental issues from a broad development perspective." (Government bill 1989/90:100; 5 p. 59.) The need for additional resource and technology transfers from the developed countries was especially pointed out. As was the case with the Brundtland report, though, these thoughts did not seem to have any affects on the Swedish policy choices vis-à-vis the Baltic Sea region.

The absence of environmental considerations in the early support program to the Baltic region probably reflected two things. First, the main goal with the program was to support the sovereignty of the new states. Although this was not explicitly stated from the beginning, the later importance given to this purpose suggests that it had at least some relevance from the beginning. This kind of support had its predecessors motivated on somewhat dissimilar grounds, but they all intended to further Swedish interests. As mentioned in the previous chapter, Sweden actively supported the Allende regime in Chile, the Sandinista movement in Nicaragua and the ANC of South Africa, to mention a few controversial examples. These support programs were not solely focused on the traditional attempt to reduce poverty and to reach people with the most urgent needs.

Second, the political changes in the former Soviet Union during the turn of the decade took place at a very fast pace. The extraordinary political turmoil made it hard to see how best to frame the Swedish support. One of the results was that the level of assistance was specified first, followed by the allocation of these resources. The Swedish government obviously was of the opinion that some kind of substantial assistance was appropriate, even though it was not clear exactly what objectives it should fulfill.

It seems clear that the formulation of the Baltic Sea initiative was held almost separately from the more general, and traditional, Swedish development policy. This was somewhat surprising, as both resource flows emanated from the development budget. As mentioned above, 90 percent of the Baltic Sea Billion was taken from the regular aid budget. The dichotomization into, on the one hand, traditional aid and on the other hand the new assistance to the CEE countries described in the previous chapter can, however, explain that aid resources were used for projects targeted towards mutual gains. More "advanced" recipient countries were selected, using a different set of criteria from those applied to traditional Swedish aid recipients. Criteria which made it possible to refer to rationales such as support of policy changes in desirable directions, mutual needs to combat transboundary environmental threats, national security and the like. This part of the assistance was to be primarily administered by BITS, which was the most experienced aid agency when it came to cooperation with more advanced counterparts.

3.1. Mutual gains and support for the democratization process

Swedish support to Central and Eastern Europe began to take shape during 1990, and became institutionalized to a higher degree than before. The so-called *Eastern Europe Report* (Östeuropautredningen), carried out within the Cabinet, argued for an increased Swedish presence especially in the Baltic States. The enlarged allocation of resources to Central and Eastern Europe in general and the Baltic republics in particular, meant that less was available for other regions, which was explicitly expressed in the Government (bill 1990/91:100:5 p. 7). As for multilateral support, an additional amount of 500 million SEK was allocated to the European Bank of Reconstruction and Development (EBRD). The Swedish government endorsed all activities taking place within this issue-area in for example the European Council, the UN Economic Commission of Europe (ECE), and the Conference on Security and Cooperation in Europe (CSCE).[86] In 1990, the parliament voted for the establishment of a joint Nordic, environmental financial institution, the Nordic Environment Finance Corporation (NEFCO). Its mandate was to finance investments in environmental equipment and services. The Swedish contribution was approximately 110 million SEK during the following six year period.

The general trend of increased attention given to trade issues and mutual benefits continued. Additional resources for the support of private business in the transforming countries were also envisioned. A new administrative body for the promotion of Swedish export interests was proposed in the parliamentary report on Organization and Procedures in Bilateral Assistance (SOU 1990:17), and this recommendation was endorsed in the Government bill. The idea was to merge SWEDFUND, IMPOD (whose task was to promote Swedish imports from developing countries) and some parts of the SIDA Industrial Department into a single institution. This amalgamation was subsequently implemented, and the result was the establishment of a new administrative body, SWEDECORP, in July 1991.

The focus was now explicitly put on furthering the development towards democracy in Central and Eastern Europe as the primary aim of the support program. This was not, however, a specific feature of the new program directed towards the CEE countries. In Sweden, as in many other OECD-countries, there were signs of increased attention being given to support of democratic changes in recipient countries. It was emphasized in the government 1990 budget bill, that Swedish authorities "consistently convey our view on democracy and human rights" (1990/91:5, p. 57). Towards this end, a new appropriation category in the Swedish aid budget was instituted, exclusively handling support for democracy, human rights and humanitarian aid. Since the Baltic States were striving for the consolidation of their democratic systems, the relevance for the position taken vis-à-vis the CEE countries was high.

As the content in terms of specific aid projects was not specified beforehand, many different administrative bodies and departments usually became involved: "In

[86] CSCE was later renamed OSCE, Organization for Security and Cooperation in Europe.

one way or another, most departments and many regulatory bodies are involved in the transfer of Swedish knowledge of administration." (Government bill 1990/91:100; 5 p. 6)

The main Swedish goal was to nourish the transformation process towards democracy, but how this should be operationalized could be interpreted in a multitude of ways. The government for example saw support of private corporations as relevant, since it could be argued that a well functioning market was a prerequisite for democratization and development (bill 1990/91:100; 5 p. 19). Strengthening of the economy in the Baltic States was also seen as a goal in itself, as they may "become important trade partners" (bill 1990/91:100; 5 p. 19).

The aim of bringing the transforming countries into the global, as well as European, economic cooperation was stated as a "Swedish goal". The notion of potential mutual national security benefits from this integrative objective was not explicitly stated in the Government bill, but still quite difficult not to concede.

3.2. Environmental support

In the 1990 Government bill, environmental support was for the first time explicitly recognized as an important part of the assistance package to the CEE countries. Of the 900 million SEK taken from the development aid budget, 400 million were earmarked for environmental assistance. More than half of this amount, however, was allotted to Poland, where environmental cooperation projects had already been initialized in 1989. During the three year program approximately 200 million SEK were directed to the Baltic States. How to allocate these resources more precisely was at that time rather uncertain, due to the unclear political situation in the region. Technical assistance and pre-feasibility studies related to waste water treatment plants in Estonia and Lithuania were mentioned, as well as environmental monitoring/analysis and energy saving projects.

In 1990, a regional conference on the environmental state of the Baltic Sea was held in Ronneby, Sweden. All the countries adjacent to the Baltic Sea were represented. This was an important occasion, since it was the first, important reflection of the perceived increase in the potential for joint action on a region-wide basis. The Ronneby Declaration and the new initiatives taken in the HELCOM process were explicitly brought into the discussion on the CEE support program.

Of interest, and in line with the description in the previous chapter of the dichotomization of Swedish aid policy, was the formulation of the purpose of the environmental assistance. The purpose was *not* to support, for example, environmental investments directly, but rather to make the countries in transition more concerned themselves about the environmental situation in their countries. The following passage is elusive:

> It is furthermore a Swedish goal to contribute to an increased level of awareness of the serious environmental problems in Central and Eastern Europe. These threaten the health conditions of the people and agricultural production within large areas. Air and water pollution also affects the environment of Sweden (Government bill 1990/91:100; 5 p. 19).

To sum up on the early phase of Swedish assistance to the CEE countries, it is clear that the approach adopted had more in common with transnational cooperation then traditional foreign aid. From the outset of the bilateral support program, cooperation and mutual benefits were envisioned. The implementing agency of most importance in this issue-area was not SIDA, but BITS, thus emphasizing the "advanced" character of the assistance. 3.3. Reactions from the political opposition

Judging by the opinions expressed by the political parties in the parliament in opposition to the government, the CEE program generally got a positive response. No party argued for reductions in the amount of support given. The Green Party (Miljöpartiet), the Center Party (Centern) and the Liberals (Folkpartiet) appeared to be most in favor of giving additional support in general and environmental aid in particular. The response by the Green Party to the CEE program was that the initiative was price-worthy, but that it should be considerably enlarged. Instead of the annual budget of approximately 350 million SEK, one billion should be reserved *yearly*, and placed into a new European Environmental Fund (Member bill 1990/91:Jo 752 p. 8).[87] The Center Party proposed a doubling of the annual 300 million SEK envisioned in the Government bill: "In order to really achieve improvements the environmental aspects have to be prioritized… and a new 3-year plan be prepared." (Member bill 1990/91:U622 p. 1) The position of the Liberals was that 50 million SEK should be added to the budget (Member bill 1990:91:U203 p. 7). This addendum should be earmarked for the Baltic States. Although the Liberals emphasized institutional and market structure support, a special funding for the SNV for this issue-area was also part of the Liberals' proposal (Member bill 1990:91:U203 p. 6).

Somewhat surprisingly, the Left Party and the conservative party Moderaterna (from here on labeled the Conservatives) both agreed with the government on the level of commitment. It should be noticed, however, that the strategies of all the opposition parties might not have been very well worked out at this point. In the case of the Left Party, this was explicitly stated in Member bills 1990/91:U516 and 1990/91:U244.

The support program to the Baltic States as initiated by the SAP government was met with nearly unanimous endorsement in the parliament, a rare response in the Swedish foreign aid debate (Standing Committee on Foreign Affairs Report 1990/91:UU02). None of the parties in the parliament questioned the need for an aid program per se. Nor was the government policy to prioritize the region closest to Sweden called into question. From the documentation of the parliamentary debate of this time, it seems almost as if the selection of the countries closest to Sweden was a self-evident choice (Standing Committee on Foreign Affairs Report 1990/91:UU02).

Apart from the different views on the volume of the package referred to above,

[87] It is somewhat unclear from the bill how much should be drawn from the European Fund for the benefit of the Baltic States, Poland and Russia. It was stated, however, that the amount should be "considerably higher" than the 300 million SEK per year suggested by the government (Member bill 1990/91:Jo 752 p. 8).

there were, however, strongly divergent opinions on how to finance the program. Hitherto, the financial resources had been taken from the regular aid budget. This was heavily criticized. The only party arguing clearly in favor of keeping the support within the aid budget was the Conservative Party (Member bill 1990/91:U531). In similarity with the position taken by the Liberals, the positive effects from increased trade and free market reforms were emphasized.[88] The Center Party had, according to their own description "... fiercely criticized the fact that the poorest people are financing activities in the Eastern Europe countries, since the resources are taken from the regular aid budget." In a similar vein, the opinion of the Green Party was that "...money allocated for environmental assistance in Eastern Europe should not be taken from the aid budget, which should be reserved for the poorest countries of the world, but should come from another budget entry." (Member bill 1990/91:Jo752 p. 8) The Left party also proposed that the financing of the support to Eastern Europe should not be taken from the regular aid budget (Member bill 1990/91:U244). No clear referrals were made concerning the weighing of interests between the traditional recipient countries of Swedish aid and the EEC. As the Left Party argued for a return to the one percent target of total aid, a reasonable interpretation is that LDCs should not receive less Swedish aid due to the CEE initiative.

The views of the parties in opposition reflected the tension between different opinions on how Swedish aid resources should be used as described in the previous chapter. Opinions differed between assisting the poorest LDCs with traditional aid for combating poverty and the redirection of part of the support to a selection of middle-income countries. There was a clear reluctance to take regular aid resources for the financing of the CEE initiative. The proposed solution was similar to the earlier split between altruistic aid and mutual gains, but this time the cleavage between the two objectives was even deeper. The support of the transformation process in the former socialist countries should not be allowed to influence the aid budget at all, but be financed from other sources. If, however, the financing of the Baltic support was to be removed from the aid budget, the implication that traditional Swedish aid criteria would become less relevant seems to have been appreciated and generally endorsed. As mentioned above, it seemed almost axiomatic that if Sweden was to give assistance to the countries in transformation, it should be directed towards the countries closest to Sweden.

4. THE POLICY OF THE NEW, NON-SOCIALIST GOVERNMENT

In the 1991 parliamentary elections, the SAP government was replaced by a coalition of four non-socialist parties: the Conservatives, the Liberals, the Center Party, and the Christian Democrats. The new government continued the support program to the countries in transition along similar lines as before, but the ambition to make a clear distinction between traditional aid and cooperation based on mutual gains be-

---

[88] See also Commerce Committee report 1990/91:NU36.

came even more marked. The new government also modified the Baltic aid in the direction suggested in their earlier critique of the former SAP regime.

The first step was to remove the financing of the CEE support from the aid budget to a new category (Government bill 1991/92:100:4 p. 8). As the regular aid volume was not decreased as a consequence of this removal, this meant an actual increase in the aid given to the traditional recipients of Swedish aid.

The second step was to significantly increase the volume of the support to the former socialist countries. In fact, even before the budget bill was presented to the parliament, the Standing Committee of Foreign Affairs issued a proposal to the parliament to make some additional 160 million SEK available for support directed to the Baltic States (Standing Committee of Foreign Affairs Report 1991/92:UU17). These resources had earlier been earmarked for an EFTA development fund for Yugoslavia, but as the Swedish financing of this fund was never ratified, the Committee recommended that its content could be redirected for other purposes. The former one billion SEK of the CEE program had been disbursed over a period of three years, while the new billion should be spent annually (Government bill 1991/92:100:4 p. 247). It should be noticed though, that some rather large items related to the support of the countries in transition were not included in the original SAP CEE program. Taken together, the initial support amounted to nearly two billion SEK (2/3 of a billion annually), which should be compared to the annual billion, set aside by the new government (see also Member bill 1991/92:U507).

The political turmoil in the former Eastern Europe had by this time become less acute, thus making it easier to allocate the support resources more consistently. The former selection of primary recipient countries of Swedish support now became more distinct.

> The support and cooperation should primarily be directed to Estonia, Latvia and Lithuania, but also to Poland and the north-western parts of Russia. There Sweden, as well as the other Nordic countries, for geographical and historical reasons has a special role to play... To the other countries of Central and Eastern Europe, Swedish assistance will be given primarily through multilateral organizations. (Government bill 1991/92:100:4 p. 7)[89]

The aim with the support to the Baltic States was according to the new government, twofold: to assist the former socialist countries in their building of democratic institutions and market structures, and to further the "significant Swedish interests concerning foreign policy, security, business and commercial issues." (Government bill 1991/92:100:4 p. 242) The government view of these twin goals seemed to be that they were fully compatible, possibly even having synergistic effects on each other. By supporting the build-up of stable, free market-oriented democracies in the former Soviet Baltic region, interests vital to Sweden would be served. Transfer of

[89] This policy to prioritize countries and regions in relative proximity was not a particular Swedish feature. In fact, one of the Swedish arguments behind the favorable treatment of the Baltic States was that since the donor countries in Central Europe tended to assist the countries closest to them, it seemed only reasonable to primarily support those closest to Sweden.

Swedish know-how and expertise was seen as the main vehicle towards this goal. Different kinds of cooperative efforts should, according to the government, be initiated on a broad front, engaging local as well as national actors and non-profit as well as business organizations. The idea seems to have been to institutionalize a spacious web of interactions between Sweden and the former socialist countries across the sea, in order to pull them closer to the Western world in general and to Sweden in particular. If this attempt was successful, Swedish security and business interests would be considerably augmented at the same time as the explicitly stated objectives of the Baltic States were supported. In other words, the Swedish support program would meet the mutual gains criteria. Swedish support for environmental investments or other means to remedy environmental disturbances was, however, not yet envisioned as an important part of the assistance package. A few referrals to this kind of support were made in a rather casual way in the 1990/91 budget bill, but were not further elaborated.

The response from the other parties in the parliament to the new government's policy vis-à-vis Eastern Europe was generally positive. The SAP and the Left Party endorsed both the removal of the assistance from the regular aid budget and the increase in volume to one billion SEK. The earlier critique of the SAP government for using aid money to finance the assistance to Eastern Europe was not valid, according to later comments by the SAP. In a statement in Member bill 1991/92:U507, aid money was used "since the budget had already been put together, and that a temporary solution to the financing problem had to be found quickly." All parties in the parliament now shared the view that the support of the countries in transition should be financed by means other than these from the regular aid budget. Furthermore, the SAP also approved of the selection of countries adjacent to the Baltic Sea as the prime recipients of Swedish aid. The same may be said concerning the ambition to involve many different kinds of actors in the interaction process, state authorities as well as municipalities, non-profit organizations, and private entrepreneurs.

In Member bill 1991/92:U507 from the SAP, the intention to make environmental support an important part of the assistance package was for the first time elaborated. The following quotation makes it quite clear that according to the SAP, environmental investments in the transforming countries would benefit the whole region:

> The serious environmental deterioration in Eastern Europe affects us directly. Our common sea, the Baltic Sea, continues to be polluted, primarily by emissions from the Eastern European countries. The common air space continues to bring substances to us, which acidify our forests and soils. We still have a long way to go before we have transformed our own methods of production, but if we cannot formulate a common environmental policy for the whole region with our neighboring countries, the environment will continue to deteriorate also in the times to come. (1991/92:U507 p. 1)

With the benefit of hindsight, it is interesting to see that some of the large investment programs subsequently initiated had been proposed quite early. In Member bill 1991/92:Jo668 by two representatives of New Democracy, for example, it was stated that Sweden ought to invest where "... we get the most out of our investments", and consequently choose projects in countries on the other side of the Baltic

Sea.

In a similar vein, a few representatives of the Liberals in Member bill 1991/92:U522 suggested that: "Better and more efficient waste water treatment plants should be built." Security issues concerning the nuclear power plants in the Baltic States were also proposed, as important targets of Swedish assistance. The need for regional cooperation was further emphasized "... in order to reduce emissions to the Baltic Sea."

Finally, in Member bill 1991/92:U505 signed by two members of the Conservative Party, explicit referrals were made to the practice of tying Swedish environmental assistance. It was stated that it would be reasonable to use Swedish expertise and knowledge for the assistance of the Baltic States, where purification plants treating emissions to air and water were mentioned as a specific area where Sweden could make important contributions.

These proposals, however, did not gather any widespread support at that time. Rather, they reflected opinions of limited groups and single party members.

5. THE EUROPEAN IMPERATIVE

In early 1993, Sweden formally declared its intention to join the European Union. The Swedish move towards Europe was reflected in many different ways. Not only was it perceived as important to Sweden to become more closely associated with continental Europe, but also to support the former socialist countries in their rapprochement to the Western world in general, and Europe in particular. It was stated, for example, in the Government bill 1992/93:100:4 that it was important to extend the CSCE (Conference on Security and Cooperation in Europe) to also include Central and Eastern Europe. To promote the development of democratic institutions in the countries in transition, Sweden espoused membership in the European Council for the Baltic Republics (Government bill 1992/93:100:4 p. 2). Sweden signed free trade agreements with Estonia, Latvia and Lithuania on a bilateral basis. Declarations of cooperation were signed between the EFTA countries and the Baltic States. At the multilateral level, the IMF, IBRD and EBRD were all intimately involved in the financing and preparation of different kinds of projects.

Although the primary motivation behind the Swedish support to the Baltic States and the Russian St. Petersburg region was security concerns, the environmental parts of the support package were now becoming more elaborated. The potential tension between self-interest and altruism was de-emphasized by the government:

> To promote the development of stable democracies, an economic system based on free market principles and an enhancement of the environment in these [Central and Eastern European] countries is of utmost importance both for them and for the whole of Europe. At the same time as it is our moral duty to assist this burdened people; it is also in our and other rich countries own interest. I therefore find it of significant importance that we, in spite of the very strained economic situation, continue the proposed three-year program with an unaltered level of ambition. (Government bill 1992/93:100:4 p. 23)

A decreasing level of centralization now characterized the approach. It was explicitly stated, "a responsibility for the protection and enhancement of the environment shall permeate the complete Swedish program of cooperation." (Government bill 1992/93:100:4 p. 26)

6. THE SWEDISH CLIMATE STRATEGY

The structure of the energy taxes in Sweden was revised in 1991. Part of the increased revenues from these revised tariffs was used for certain measures to improve the environment in the CEE countries. No less than 500 million SEK were received for the development of renewable energy sources and energy efficiency investments in Government bill 1991/92:150. One year later, Sweden ratified the UN Climate Convention in Rio de Janeiro (1992), and in the second half of the same year, measures to reduce carbon dioxide emissions were outlined in Government bill 1992/93:99 and 179. These factors were all important in the process of formulating the Swedish Climate Strategy.

The initiatives in the issue-area of abatement of greenhouse gases became important for the evolution of the Swedish approach vis-à-vis the countries in transition adjacent to the Baltic Sea. There were two key standpoints adopted in the Swedish greenhouse gases strategy, with high relevance to the cooperation with the countries in transition. The first concerned efficiency. The first step is to find the substances most detrimental to the atmospheric climate. Then the cost of reductions in different areas should be compared to each other. Different kinds of emissions should then be decreased according to their cost-benefit ratios. In Government bill 1992/93:99:1 it was established that:

> The order of preference between measures to reduce emissions of carbon dioxide and other greenhouse gases should be decided according to their importance on affecting the climate *in relation* to the costs. (Author's italics)

The other key position adopted in the Swedish climate strategy concerned the permeability of national borders. As the atmosphere in most practical aspects is, if not a global common, at least a regional one, it is of limited importance to reduce emissions in a single, small country. This was reflected in the same Government bill:

> An effort to make the energy supply in the Baltic Republics more efficient and based on renewable sources would have several important advantages. It should among other things be one of the most cost efficient measures Sweden could undertake in order to reduce global carbon dioxide emissions.

From the 500 million SEK allocated to finance the Swedish climate strategy in 1993,

45 million were earmarked for investments in the three Baltic Republics.[90] The objective was to get a maximal decrease of greenhouse gases from a given amount of money, irrespective of where these investments were implemented. Since the net cost of these investments was lower in the Baltic States than in Sweden, the money ought to be invested there. This line of reasoning seems sensible, given the aim to reduce emissions of greenhouse gases. It does, however, also clearly reflect a cooperative approach, rather than a primary ambition to assist the Baltic States. Sweden was committed to decrease emissions of greenhouse gases. If the prime purpose had been to assist the Baltic States, the most urgent environmental needs in these countries should instead have been the basis for the allocation decision.

Within the budget for the Swedish Climate strategy, the environment of the Baltic Sea was also given increased importance. At the Ronneby conference in September 1990 mentioned above, it was decided by the littoral states that a working group should be set up under the auspices of the Helsinki Commission, assigned to the purpose of the preparation of an action-plan (HELCOM 1995). This action-plan should be directed towards the goal of a comprehensive restoration of the Baltic Sea.[91] The Environmental Ministers of the littoral states endorsed the "strategic content" of the plan in April 1992 (Government bill 1992/93:99:1 p. 6). It was suggested by the Swedish government that an amount of 142 million SEK should be directed to measures defined as urgent in the preliminary action-plan (Government bill 1992/93:99:1). The funding of the Climate Strategy should finance this.

7. SAP APPROVAL OF THE GOVERNMENT STRATEGY

The approach adopted by the government vis-à-vis the countries in transition was, by and large, approved by the political parties in opposition. The SAP strongly criticized the propositions of the government, but when comparing the actual proposals, the differences between the position of the government and that of the SAP were not very significant (Party bills 1992/92:U251, Jo14 and U409). The government was primarily blamed for not having elaborated a comprehensive strategy in this issue-area.[92] Too much was left for ad-hoc adaptation and incremental problem solving. This critique, however, seems to primarily have been a reflection of the difference between being in government and being in opposition. As described above, the SAP government was criticized for similar shortcomings during their time in office. The

---

[90] Whether investments for the reduction of greenhouse gases financed by one country but implemented in another has been intensively discussed ever since the adoption of the Convention. In Article 3.1 it is stated that the climate system should be protected "...on the basis of equity and in accordance with their common but differentiated responsibilities and respective capabilities." It is thus clear that efforts to reduce emissions may vary depending on both current levels ("responsibility") and ability to afford such efforts ("capability"). The Convention is silent, however, on how transnational initiatives should be reflected in the national accords of emission reductions. For a discussion of the approach chosen by the Nordic countries, see Ringius, Torvanger & Mezears (1996).

[91] The Baltic Sea Joint Comprehensive Environmental Action Program will be discussed further below.

[92] The Standing Committee on Foreign Affairs also expressed similar opinions regarding the need for a comprehensive assessment of the Swedish assistance to Central and Eastern Europe (1992/93:UU16).

lack of a comprehensive strategy vis-à-vis the countries in transition on both occasions reflected the inherent difficulties of formulating such a strategy. The rapidly changing political situation and a shortage of administrative capacity were more important factors behind the lack of a comprehensive strategy than diverging views on the preferred content of the support.

Concerning the prioritized areas, the SAP endorsed the government preference given to the Baltic States, as the primary beneficiaries of Swedish assistance in the Central and Eastern Europe program. In addition to this, special interest should be paid to the St. Petersburg region and Poland, something that was not argued against in the government proposition. The level of funding of BITS was described by the SAP as a reflection of an "aversive attitude" by the new, non-socialist government (Party bill 1992/93:U251 p. 3). In actual figures, however, the SAP proposal was only to raise the level from 200 million SEK annually to 250 million. The frame proposed by the government for export credit guarantees, one billion SEK, was endorsed fully by the SAP.

The almost unanimous perception of mutual interests and the potential benefits from increased trade between Sweden and the Baltic States was striking. All political parties in the parliament were very explicit in their statements concerning the benefits from support directed to market reforms and international trade. One of the most out-spoken parties on this issue was the Left Party. In Party bill 1992/93:U258 on "Cooperation with Central and Eastern Europe" the views expressed by one of the most famous businessmen in Sweden, Mr. Percy Barnevik, head of the ABB, were extensively referred to. According to Mr. Barnevik "...if there were 200 Western European companies investing as much in Eastern Europe [as ABB has done], it would create 5 million jobs directly, and 10-20 million indirectly, by stimulating subcontractors, transport providers etc."[93] According to the Left Party "... the Swedish state has a responsibility to stimulate such investments. Investments which would create work opportunities also in Sweden." (Party bill 1992/93:U258 p. 2) The views of Mr. Barnevik were not very surprising. What was more unexpected was the degree of convergence on this issue between a spokesman for large business interests and the socialist Left Party. A large part of the envisioned investments should, according to the Left Party be directed to the environmental sector, e.g. energy savings and reductions of harmful emissions. The rationality behind the government strategy to invest where the largest environmental effect in relation to its cost was reached, was completely agreed upon by the Left Party:

> Massive investments in infrastructure would decrease energy use and increase environmental concern, both of which are for the benefit of sustainable development. They also stimulate Swedish industrial capacity and job creation. The energy sector is of primary relevance, since implemented investments in this sector give a return many times the original investment in terms of environmental gains and energy savings. (Party bill 1992/93:U258 p. 3)

---

[93] These views of Mr. Barnevik were presented in an interview in the Swedish newspaper *Svenska Dagladet* (November 3, 1992).

Only the Center Party expressed some reluctance concerning the rationality in prioritizing environmental investments in the countries in transition rather than in Sweden. The Center Party had the longest record of placing environmental issues at the top of the political agenda and was a member of the coalition government. In Party bill 1992/93:Jo631 p. 2 it was stated, "it is important that this policy to invest where the marginal environmental utility is largest does not mean that fewer resources should be available for investments in Sweden." The only party with a somewhat divergent view on the over-all strategy of assistance to the countries in transition was the political party New Democracy, a party elected to the parliament for the first time in 1991. They did not question the existence of the support program. On the contrary, they wanted to see it doubled, in terms of financial resources. But they differed from the other parties in their opinion on how to finance the program. According to the New Democracy Party, Swedish foreign aid policy should be changed in several ways, and an increased emphasis should be put on regions close to Sweden. The Swedish support to the countries in transition should, reflecting this change of focus, be financed over the regular aid budget and be explicitly tied to Swedish goods and services (Party bill 1992/93:U255 and U219).

8. THE REVISED ASSISTANCE PROGRAM

In the 1994 government budget proposal, the need for an evaluation of the experiences of the support program and an elaboration of the future strategy was appreciated. At least three important reasons for this were discernible in the Government bill.[94] First, the political situation in the Baltic States and Russia had become more stable, in comparison with the first years of the 1990s. Estonia, Latvia and Lithuania were now all formally sovereign nations. Consequently, the Swedish sovereignty support to these countries from now on would be aimed at the further institutionalization of a free market and democratic institutions. Second, Swedish negotiations to become a member of the European Union (EU) had formally begun in February 1993, and had now entered a very intensive phase. Sweden was not the only applicant country in this process, but was joined by Norway, Finland and Austria.[95] As a joint Nordic membership in the EU might have resulted in greater Swedish influence than if Sweden had joined the Union alone, it was seen as important by the Swedish government to carefully assess the impact of membership, not only for Sweden but for the region of Northern Europe as a whole.

---

[94] See also the report on the future organization of the Swedish cooperation with the CEE countries in Ds 1993:79.

[95] After referendums held in respective countries, Norway, however, chose not to apply for membership.

> Our policy vis-à-vis Europe aims towards placing Sweden at the core of European co-operation. For us, it is of vital importance that the present cooperative activities are broadened and made deeper. Sweden should exert a significant influence on the formulation of the answers of the most important issues of the 1990s; peace in Europe, economic growth, employment and welfare, the cooperation with the countries in Central and Eastern Europe, and transboundary pollution (Government bill 1993/94:100:4 p. 5).

Third, Swedish implementing agencies and authorities of the support program had by now had some years of experience. The knowledge they had gained through the first three years of the CEE program was dispersed, and compilation and evaluation was needed in order to formulate the strategy for the years to come. As the first three-year program would end in June 1995, a potential renewal and reformulation strategy for the new three-year program had to be initiated in 1994.

In June 1993, a one-man commission had already been set up by the government, with the assignment to compile gained experiences from relevant authorities and agencies and to make suggestions for a more coordinated future strategy for the Swedish CEE program. The assignment was, however, rather circumscribed. No attempts to evaluate the outcomes of the assistance so far should, for example, be made. Only four months later, the brief report *Direction and Coordination of Swedish Cooperation with Central and Eastern Europe* was presented (Finansdepartementet 1993). At the administrative level, several evaluations had also been undertaken by the agencies engaged in the support program. Finally, investigations were also undertaken concerning the establishment of a "Sweden-house" (Sverigehus) in St. Petersburg, Russia, to increase Swedish presence in the area (Government bill 1993/94:100:4 p. 18).[96]

The general trend concerning the increased emphasis on free market reforms, international trade and human rights in the Swedish development policy continued in the mid-1990s. Although the original goals were still relevant, it was now also legitimate to use more far-reaching conditional demands on the recipient countries:

> In order to make the aid efficient... respect must be paid to the development of democracy, respect for human rights, and a smooth working free market..." (Government bill 1993/94:100:4 p. 5)

9. REFORMULATION OF THE CEE-ASSISTANCE GOALS

In order to give the future support somewhat sharper strategic contours, a new set of goals were formulated for the CEE assistance. The aim was now said to be (Government bill 1993/94:100:4 p. 156):

- The restoration and consolidation of democratic and judicial institutions
- To support the re-institutionalization of a free market economy

[96] The creation of a "Sweden-House" in St. Petersburg had, however, been proposed by the SAP government before in 1991.

- To support environmental abatement activities, especially concerning the Baltic Sea
- To assist the Baltic States in the consolidation of their autonomy

The Baltic States were now explicitly to be the primary beneficiaries of the CEE support program. The ambition was, however, to also involve the north-western part of Russia. In the spring of 1993, the Swedish Prime Minister visited Moscow, and signed an agreement on intensified cooperation between the two countries. Judging by the statements by the Swedish government, it was no less important to strengthen ties with Russia, although the practical difficulties were perceived as considerably larger than with the comparably small and homogenous Baltic States. The Baltic Sea Council, created in 1992, where all the littoral states were members, served as an important forum for discussion.[97] The mission of the council was not to bring up sensitive security issues, but rather to provide an opportunity to discuss day-to-day issues and low-politics, such as for example common environmental issues, communications, trade etc.

At this time, in 1993 and 1994, several experts (e.g. the Swedish economist and expert advisor to the Russian president, Mr. Anders Åslund) advocated a fast liberalization and privatization process.[98] A so-called "chock therapy" would, according to these experts, most efficiently transform the former socialist states into modern liberal democracies. According to them, the reforms had to keep the momentum in order not to be inhibited by vested interests in the bureaucracy. These recommendations were, however, criticized by the SAP (Member bill 1993/94:U808, U304). This criticism also became a critique of the government: "The non-socialistic government of Sweden has made the ideology of 'chock therapy' central to the whole Swedish East policy" (Member bill 1993/94:U808, p.2)

Although there were hardly any disagreements on the content of the support program, some differences between the non-socialist government and the SAP were discernible. For example, the SAP suggested that the funding from the Swedish climate strategy mentioned above should be merged with the regular East support budget (Member bill 1993/94:U808). By doing this, and by making a more clear demarcation between the government and its agencies, a more coherent Swedish policy would by achieved, according to the SAP. The government, however, rejected this proposal on the grounds that assistance projects should not be mixed up with investments within the Swedish climate strategy (Standing Committee on Foreign Affairs Report 1993/94:UU16). This seems to indicate that the government opinion was that the self-interest component in the Swedish support program could be further partitioned, one part being completely structured around Swedish prefer-

---

[97] The idea of creating a Baltic Sea Council was first proposed by the Green Parties in Sweden, Estonia and Finland jointly in 1990 (Member bill 1990/91:U501) It was subsequently picked up by several other parties, most notably the SAP (Member bill 1991/92:U506, 1991/92:U507).

[98] Sweden assisted Russia in the early 1990s on national policy-making through two separate projects; one administered by the Swedish Department of Finance and one by the East Economic Institution. Mr. Åslund was part of the latter initiative.

ences and one part reflecting both Swedish and recipient country preferences. It should be noticed though, that this distinction was not very clearly defined. According to the Standing Committee on Foreign Affairs it was of "... a technical nature" (Standing Committee on Foreign Affairs Report 1993/94:UU16 p. 19).

10. THE NEW SAP GOVERNMENT INTENSIFIES THE ASSISTANCE

In the parliamentary election of 1994, the SAP formed a new government, thus replacing the non-socialist coalition. In the Declaration of government, the intent to intensify cooperation with the littoral states was announced. As decided in a referendum, Sweden would become a member of the EU in January 1, 1995, and it was seen as very important by the Swedish government to bring the regional cooperation of the Baltic Sea countries into the broader European context (Government bill 1994/95:100:4).[99] This intent was not affected by the change in Cabinet.

In the summer of 1995, the on-going three-year support program would come to an end. As one of its first tasks, the SAP government began a comprehensive evaluation of the assistance and the formulation of a new program. According to the Government bill 1994/95:160, devoted entirely to Swedish cooperation with Central and Eastern Europe, a third phase of the collaboration had now commenced. In the first phase, 1988-1991, the Baltic States had regained their sovereignty. The second phase stretching from 1991 to 1994 was one of consolidation. If the Swedish goal in the first phase had been to support the sovereignty aspirations, the motivation in the second was to make the process of democratization irreversible. Now entering the third phase, the Swedish objective was to further the integration of the Baltic States into the European political sphere. "A reasonable assumption is that [this phase] will end within a decade or so, with the full membership of the Baltic States in the European Union." (Government bill 1994/95:160 p. 13)

Much in the same way as the traditional Swedish development aid imperatives earlier had been concretized in a few, general goals, the SAP government now formulated four principal objectives for the cooperation with Central and Eastern Europe (Government bill 1994/95:160 p. 5):

- Further collective security
- Deepen the culture of democracy
- Support a socially sustainable economical development
- Support an environmentally sustainable development

It should be noted, that partly in contrast to the "old" goals of Swedish foreign aid policy, these aims were clearly focused on joint benefits. Sweden had been arguing in numerous fora for the connections between social stability, democracy and collec-

[99] This ambition to put more emphasis on the Baltic Sea region was appreciated by the EU, as reflected in for example the EU Baltic Sea Strategy advanced during 1994. In addition to this, a comprehensive European strategy was adopted by the European Council at a meeting held in Essen, in December 1994.

tive security. Bearing in mind that the recipients were all neighboring countries, the implications on Swedish security were obvious. The support of for example "a socially sustainable economic development" on the one hand was clearly valid irrespective of who the recipient of the assistance was. On the other hand, it is also clear that the implications for Swedish security were quite different if social unrest should take place in neighboring countries rather than elsewhere in the southern hemisphere.

The last goal on environmentally sustainable development may seem quite similar to the one in the general formulation of Swedish foreign aid policy, except for the merging of development and environmental protection. There were, however, some differences. In January 1994, the Committee for Analysis of Development Cooperation (KAU) had been asked to, in addition to its original mission to analyze Swedish foreign aid generally, study the cooperation with Central and Eastern Europe. One of its conclusions was that the environmental goal should be more precisely formulated. The Committee proposed the following alternative formulation: The goal should be to "support environmental cooperation and procedures for the protection and enhancement of the environment, especially the Baltic Sea and the region surrounding it, and to decrease the risks for large, environmental disasters." (Utrikesdepartementet 1994b) Although this formulation was not incorporated in the goal statement of the government, it did reflect a sharpened focus on the Baltic Sea region.

11. ADMINISTRATIVE REORGANIZATION

The SAP government proposed quite large rearrangements at the organizational level. The SAP had previously criticized the plethora of institutions and governmental bodies involved in the assistance in one way or another. In Government bill 1994/95:160, it was suggested that a new, larger aid administration should be created. The three largest administrative bodies, SIDA, BITS, and SWEDECORP should be merged into one aid institution.[100] This institutional amalgamation was undertaken in the summer of 1995.[101] A special unit was created for the exclusive handling of Swedish assistance to Central and Eastern Europe, Sida-East. Sida-East was given the responsibility for the implementation of the bilateral part of Swedish support to CEE.[102] Most of the competence of the former bureau was, however, transferred to the new aid agency. According to the SAP government, this reconstruction of the administration would increase efficiency, competitive procurement, and evaluative capacity (Government bill 1994/95:160 p. 8).

---

[100] The Board for Development Studies (SAREC) and Sandö Development Center (an educational establishment for the preparation of presumptive aid workers) also became assimilated in the new Sida.

[101] The new authority kept the name Sida. The only difference from the previous name was that the acronym was no longer spelled entirely with capitals.

[102] It was, though, stated explicitly in Government bill 1996/97:1 section 7, that the implementation of the support should be carried out in close cooperation with expertise bureaus (e.g. the environmental component should be implemented in close contact with the Swedish EPA).

At a joint Nordic level, the Nordic Environment Finance Corporation (NEFCO), which had been working on a temporary basis since 1990, was now given permanent status.[103] The suggestion by the Swedish government was that it should be given a similar level of financing as during the first years of its existence.

The Swedish support to CEE countries had now started to become adapted to the different characteristics of the recipient countries, at least to some extent. The total environmental support to the CEE countries was budgeted at some 300 million SEK during the 1995/96 budget year (nuclear safety included).[104] This equaled about 30 percent of the total support. The main target sectors did not change. Most assistance was given to waste water treatment, air emission reductions and nuclear safety. Although it was stated in the 1994/95 Government bill that priority should be given to transboundary as well as local environmental hazards, there is no doubt that the main focus was put on the former. The Baltic Sea was targeted in particular. Apart from three waste water treatment plants in Poland, which in early 1994 were almost completed, three large plants were scheduled to be initiated in early 1995 in each of the Baltic States. One would be located in Klaipėda (Lithuania), one in Liepaja (Latvia), and one in Haapsalu (Estonia). Sweden was in all these cases the main sponsor, while the IBRD acted as a co-financier.

In addition to the Swedish commitment in the sector of waste water treatment, two additional prioritized areas should be mentioned. In Estonia, the large scale burning of oil shale was recognized early on as a major contributor to transboundary sulfur emissions in the region. Large investments in energy conversions were needed to reduce these emissions, which to some extent were carried by the wind to Swedish territory. In Lithuania, one of the major concerns was the Ignalina nuclear power plant.[105] Swedish experts saw the security mechanisms of the Ignalina facility as deficient. The initiatives taken by Sweden were to some extent influenced by recommendations from IAEA and the G24 Nuclear Security Coordination Unit.

The reaction from the other political parties in the parliament on the Government bill on Cooperation with the CEE countries was in general terms very sympathetic. In fact, the almost unanimous support from the opposition parties was quite remarkable. The importance, for example, of opening the markets in the Western countries for exports from the CEE countries was likewise appreciated by the Conservative Party and by the Left Party (the two parties generally regarded as being furthest apart on the traditional Left-Right dimension). The Conservative Party stated that:

---

[103] The decision to create a joint, Nordic, environmental investment institution had been taken by the Nordic Council of Ministers in March 1990.

[104] This figure was the estimation given by the government. The difficulties inherent in making such estimates should, however, be emphasized. This figure refers to projects where the *main objective* has been to achieve environmental effects. This does not mean that other projects, i.e. energy efficiency, corporate reconstruction etc, have no environmental effects. In most cases they do.

[105] Sweden took some further measures to reduce the risks connected with radioactive waste in Estonia, Latvia and the north-western part of Russia. The nuclear power plant in St. Petersburg (Sosnovij Bor) was seen as a probable target for future Swedish support.

> The most important factor in the Western support to the transformation of the countries in Central and Eastern Europe is allowing entrance to its markets (Conservative Member bill 1994/95:U30, p. 5).

The wording of the Left party was very similar:

> ...we lack ambition and visions when it comes to the most important issue: How Western countries will open their markets for competition from the CEE countries (Left Party Member bill 1994/95:U33, p. 6).

Also regarding the choice of recipient countries, the preferences of the political parties were still unanimously in support of the riparian states of the Baltic Sea (Member bill 1994/95:U808, U33, U32, U31, U30, and U26). The main portion of the resource flow should be destined for the Baltic States, the St. Petersburg region and, albeit decreasingly, to Poland. The SAP government was criticized for not paying enough attention to the cooperation with north-western Russia. The government had, however, stated that their opinion did not differ on this matter. The fact that Russia had received less aid was a reflection of practical and political difficulties, rather than lack of interest. This was not questioned in any of the Member bills. Nor was the view that it might be argued that Swedish assistance should have been targeted towards meeting the most urgent needs in Central and Eastern Europe, which might not have been located in the Baltic Sea countries.[106]

12. MINOR DIVERGENCES

There were, however, two minor divergences between the government and some of the other political parties in the parliament. The first one was related to the level of bilateral, environmental assistance, and the other to the administration of it.

The bilateral, environmental assistance to the CEE countries had decreased by about 20 percent, compared to the previous budget-year. The Center Party and the Green Party, but also the Conservatives, criticized this most vigorously.[107] This reduction was, though, not a reduction of environmental assistance, but a cutback of *bilateral* assistance. About four fifths of the cutback was instead distributed through the European Union CEE assistance administration to similar kinds of projects (Government bill 1994/95:160). The remainder of the decrease was, according to the government, merely a fluctuation caused by temporal deficiencies in the administrative capacity of the recipient countries (Committee Report 1994/95:UU16, p. 62).

Although the attempt by the government to decrease the administrative disparity of the Swedish CEE assistance was generally praised by the other parties, the merger of the aid bureaus into one "super-Sida", as labeled by the Conservatives, was not seen as a very good idea by this party and the Center Party. As the assistance to the Baltic States ought not to be regarded as a cure for "underdevelopment" but rather as

[106] Except for Disaster relief, which might be used in any of the CEE-countries if an urgent need should arise, according to the government.

[107] The Green Party proposed a doubling of the assistance to the CEE countries (Member bill 1994/95:U26). This was firmly rejected by the government (Committee Report 1994/95:UU16, p. 62).

a means of meeting the needs created by a "faulty development" in the former socialist countries, it should be handled separately.[108] According to the Conservatives, the assistance to the CEE countries should not be mixed up with traditional Swedish aid, as it should be characterized by a cooperative attitude, rather than one of aid (Member bill 1994/95:U808, U30). The Center Party was against the administrative changes in the handling of environmental assistance. The government proposal transferred the main responsibility for environmental assistance from the Environmental Department to the Foreign Department, which was seen as threatening the quality of the environmental part of the support to the CEE countries.

The divergent opinions between the government and some of the opposition parties reflected the dual aspects of Swedish assistance described previously. In this context, the issue came down to whether the CEE assistance should be regarded as foreign aid (albeit with important aspects of mutual interests) or as cooperation between two relatively equal parts. As had been the case before, the SAP was more apt to view the assistance as similar to other kinds of aid, while the non-socialist parties tended to give larger weight to the dissimilarities. The picture was, however, far from clear-cut. The main characteristic of the views of the different parties is one of great similarity. The few expressed differences were in many cases not following the traditional left-right political dimension, but had a more complicated pattern of party and individual preferences.

The comprehensive treatment of the Swedish support to the CEE countries in the special Government bill 1994/95:160 influenced the policy in this issue-area in a long-lasting manner. The changes in the content of Swedish assistance during 1996 and 1997 were all relatively small. The budget for the over-all assistance was kept at one billion SEK annually, as in the years before.[109] In addition to this, a Baltic regional fund was established in the spring of 1996. The fund budget of one billion SEK fund was to be administered by the new Baltic Council of the Prime Minister (Statsministerns Östersjöråd), and was to be used for cooperation and development in the Baltic Sea region during the following five years.[110]

13. DIVERSIFICATION AMONG THE BALTIC RECIPIENTS OF SWEDISH ASSISTANCE

The Swedish assistance had hitherto been directed to the three Baltic States on a relatively even basis. Although it had been noticed that there were differences between different countries and sub-regions, this had generally not resulted in a clear adaptation to these diverging circumstances. In the special Government bill 1994/95

---

[108] The Conservatives invented the "faulty development" expression (Member bill 1994/95:U808, U30).

[109] This amount was also stipulated in the new three-year program for the period July 1, 1995, to December 31, 1998, proposed by the government in 1994/95:160, and decided upon by the parliament in April, 1995.

[110] The Baltic Council of the Prime Minister should not be confused with the Baltic Council, which was set-up on a German-Danish initiative in 1992.

on the cooperation with Central and Eastern Europe it was proposed, however that separate strategy documents should be formulated for each recipient country. In 1996, the government decided upon such documents for each of the Baltic countries. According to this decision these strategies should govern the complete Swedish development program for each of the countries.[111] It is quite clear, however, that these strategy documents did not signal any great changes of future Swedish policy vis-à-vis the Baltic States. As had been stated many times before by the SAP as well as by the non-socialist government, the ultimate aim was not to institutionalize Swedish aid for a long period, but rather to support local institutional and administrative capacity in order to make further assistance superfluous. In practice, however, the experience has been that the transition from Swedish to local administration and financing has often been problematic, which has repeatedly prolonged the Swedish presence.

The content of the Swedish environmental aid would, according to the strategy documents, continue along the same lines as before, thus focusing primarily on improvement of waste water treatment, energy efficiency and nuclear safety. In Estonia, the mining of oil shale was confirmed as a highly prioritized area, in Latvia waste water treatment was given special attention, and, in Lithuania, safety measures at the Ignalina nuclear power plant were the largest, single target for Swedish aid.

In all three strategy documents, it was clear that one of the most important factors in the Swedish relationship with the Baltic States, was the integration of these countries in a Western European framework. Full membership in the European Union would be the ultimate reflection of this. The integration of the Baltic States in the Western sphere was seen as not only beneficial to Sweden from security and economic perspectives, but also from an environmental standpoint. As these countries would be receivers of assistance, Sweden would be able to influence the content of large-scale EU-programs (e.g. PHARE-programs). A decreased distance between the West and the Baltic States would further give Sweden increased opportunities to influence the decisions with bearing on the Baltic States taken in the large, financial institutions, such as the IBRD and the EBRD.

> Not least in the Baltic States Sweden has, due to its comprehensive cooperation, in many cases the possibility to influence the design of, for example, the programs of the IBRD and EBRD (Government bill 1996/97:1 p. 11).

As has been noted before, there were never any large and important differences between the government and the other political parties, neither when the non-socialist bloc held the cabinet, nor when the SAP governed. In fact, in 1996, it was hard to see any differences in opinion at all. The SAP had previously criticized the establishment of the business school in Riga, which had been initiated by the earlier Swedish non-socialist government.[112] Not anymore it seemed, as in the budget bill

---

[111] Government decision II:11, UD96/1834/EC (Lithuania), II:10, UD96/1835/EC (Latvia), and 25, U11 (Estonia).

[112] The business school in Riga was the Swedish contribution to Euro-faculty, a higher education enterprise initialized by the Baltic Sea Council.

of 1996/97, no traces of the earlier critique were to be found.

The only real traces of diverging views were on the level of assistance. The Left Party wanted to add another 300 million SEK and the Conservatives maintained that the Foreign Department should have the primary responsibility, rather than the aid administration.

Regarding the importance attached to free markets and liberalization, no party in the parliament showed any divergent opinions. The Left Party, which traditionally has displayed the least confidence in the virtues of free markets, seemed to embrace most of the policy recommendations as much as the other parties:

> The reform program [in Poland] commenced in 1990 with a so-called *chock therapy* – the price mechanism and the financial sector were deregulated, trade was liberalized, the local currency was made convertible, while OECD guaranteed the exchange rate. Over night, almost all shortages evaporated and a free market was released (Member bill 1996/97:U814, p. 3).

At the other end of the political scale, the Conservatives were strongly convinced that relatively comprehensive Swedish assistance efforts were also needed in the foreseeable future. Although they emphasized that the ultimate aim with the Swedish assistance was to make it redundant,

> [environmental] support efforts would be necessary for a long time to come in order to help these countries to get rid of their environmental inheritance (Member bill 1996/97:U813, p. 3).

The focus placed by the government on waste water treatment, energy efficiency, and nuclear safety was not questioned. Neither was the almost exclusive priority given to the countries closest to Sweden.

14. CONCLUSION

The consensus among the political parties on the existence and the character of the Swedish assistance program was quite remarkable during the entire 1991-1996 period. When the SAP government initialized the program, no other political party had any objections about its creation. The only significant issue where the opinions diverged was on how to finance the program. At first, the resources were taken from the regular aid budget, which was criticized by the non-socialist parties in opposition. Somewhat later, the financing of the CEE assistance was removed from the regular aid budget, and placed in a new category in the budget bill. This was never really disputed by the SAP.

The recipients of Swedish aid have almost exclusively been the countries closest to Sweden. Poland was the largest recipient in the early years, but as the Polish economy grew, this support was scaled down in favor of the Baltic countries. None of this was disputed by any of the political parties in opposition. There seems to have been a widespread consensus on the preference given to the Baltic Sea countries rather than according to a strict ranking of needs.

The Swedish support was almost entirely targeted to upgrades in the waste water treatment and energy sectors, both of great importance in relation to transnational pollution. The practice of using tied aid was never really called into question. On the contrary, it was emphasized by almost all members of the parliament that by using Swedish resources, the support program would also benefit Swedish interests.

It should be noted that the government alternated twice between the SAP and the non-socialist coalition in the period studied. The support program was first formulated by the SAP government, which was replaced by the non-socialist coalition in the 1991 election. In 1994, the SAP came back into power and formed a new government. The absence of any important differences in views on the CEE assistance over the whole period shows that both governments, while they were in office, perceived both the need and character of the support program in similar ways.

In the next chapter, the national environmental priorities of the Baltic States will be described. This description will then be used to compare the profile of the Swedish support portfolio with how the Baltic countries perceived their most urgent environmental needs.

CHAPTER 5

PRIORITIZED ENVIRONMENTAL ISSUE-AREAS IN ESTONIA, LATVIA, AND LITHUANIA

1. FOCUS ON THE SWEDISH STRATEGY

In the previous chapter, a fairly comprehensive treatment of the evolution of the Swedish environmental aid to the Baltic States was presented. The primary reason for the quite detailed description was the importance of showing that the unitary actor assumption is reasonable. For the purpose of this study, the need for a detailed description of the environmental policy preferences of the Baltic States is not so great. The primary reason is that this study is focused on Swedish behavior. Although the environmental preferences of the Baltic States are essential parts of a description of this behavior, the focus is nevertheless set on Swedish behavior. In addition to this, the relation between the donor country Sweden and the recipient countries of Estonia, Latvia, and Lithuania is of a very unequal character. Sweden is clearly in a position to formulate the content and size of the support program, while the Baltic States have more of a "take-it-or-leave-it"-decision to make.

Although it is true that the first project initiative usually came from the authorities in the potential recipient country, the gate-keeping power of the Swedish authorities and their ability to redesign the proposals means that Sweden had the effective power to draft project outlines. Somewhat simplified, it was then up to the potential recipient of the assistance to accept or reject the proposal. Despite the lesser need for a detailed description of the contemporaneous environmental priorities in the Baltic States, a fairly lengthy account will be presented below, since there is a large number of issue-areas of relevance to environmental protection. In order to give a sufficient account, it is thus necessary to comment on many kinds of environmental issues in this section. The primary objective is to identify the most important environmental issue-areas, as perceived by the Baltic States themselves. This agenda of prioritized issue-areas will then be used to analyze potential divergences compared with the Swedish support project portfolio in the analysis in the last chapter.

2. BACKGROUND

It appears as if the environmental agendas of the Baltic States in their first years of independence were very similar. Although some issue-areas were particularly salient to an individual country, the influence from the Western environmental discourse affected all countries in similar ways. Before independence, environmental priorities were made in Moscow. These were set not so much with reference to the conditions in the Baltic States, as to the needs of the central government (Dragunas 1995). This meant that when independence was achieved, the environmental strategies had to be formulated almost from scratch. At that time, it was quite natural for the re-established Baltic States to seek models and inspiration from the relatively affluent West. This resulted in similar policy choices in the three countries, despite diminishing cooperation in terms of personal contacts and institutional networks between them after their first years of independence (Vihalemm 2000).

When the Baltic States became independent in the early 1990s, they all faced severe socioeconomic hardships. Production plummeted, unemployment increased rapidly, and their trade patterns changed fundamentally. The only way forward seemed to be to try to become an integral part of Western Europe in general, and members of the EU in particular. This was a conclusion drawn by all three Baltic States. By the mid-1990s, Estonia, Latvia, and Lithuania had all applied for membership and signed accession agreements with the European Union. In order to become members of the EU, the Baltic States were required to begin a process of approaching all relevant EU regulations, directives, and legislation. The goal of this process was an eventual membership in the EU. Due to the great complexity in harmonizing the entire legal and administrative framework from the former Moscow-controlled administration this process was, however, expected to be completed gradually over a relatively long period of time (Hägerhäll et al 1996). The three Baltic States all attempted to implement market-oriented and open border policies in order to end the economic recession and to adapt to mainstream policies in the West. Of the three countries, Estonia liberalized most vigorously and had by the middle of the 1990s one of the most market-oriented economies in Europe (OECD 1996). In Latvia, the privatization of the large state-owned companies was fiercely debated (ECE 1999b), while the Lithuanian commitment to such reforms has been described as "indecisive" (Lainela 2000:204).

In this chapter, the sociopolitical context in the Baltic States and the environmental strategies formulated in the early 1990s will first be described. The environmental issue-areas perceived as most important by the national governments are then categorized. These categories are rather broad and cover wide-ranging prioritized environmental areas. The respective countries did not explicitly compare these different areas with each other with regard to their relative importance, based on for example cost/benefit ratios. This is not an observation valid only in the Baltic States, but was largely a general feature of contemporaneous environmental policies in most countries. Although it would be more efficient to invest in areas where the marginal net cost is lowest, this approach presupposes that different kinds of environmental benefits can be compared and measured against one another. Since this is problematic, the most common approach used by the authorities was to identify important issue-areas, and then to try to enhance the environmental situation in these areas.[113] Therefore, a comparison of the relative importance of the prioritized issue-areas in the Baltic States will not be attempted. The most important issue-areas are instead identified, making up a list of the most important environmental issue-areas, but with no ranking order between them. This list will be compared with the allocation of Swedish environmental assistance in Chapter seven.

In Estonia, the Supreme Council had already assessed the environmental state in 1989, and a general Policy on Nature Conservation and Sustainable Use of Natural Resources was adopted. The political reforms since regained independence did not "alter these goals, but it changed the conditions for achieving these goals." (Estonian Environment Information Centre (EEIC) 1996:6) In February 1995, the Act on Sustainable Development was adopted by the parliament. The National Environmental Strategy document was accepted by the parliament in March 1997, and was followed about a year later by the National Environmental Action Plan. As in Latvia and Lithuania, environmental activism was intertwined with the struggle for independence. In Tallinn, during demonstrations for independence from the Soviet Union in 1989, banners with the text "Baltic Sea, Our Common Sea" were carried by some of the demonstrators.[114]

The approach adopted in the Estonian Environmental Strategy document of 1996 was decidedly influenced by contemporaneous Western judicial debate.[115] It was perceived as a main principle to base the strategy on the historical traditions of Estonia and the present socioeconomic situation, but also on "…internationally accepted principles." (EEIC 1996:9) Interestingly, it was clearly stated that the environmental considerations should not be suppressed because of other societal needs:

> The interests of the environment have to stand higher than party politics, current economic problems, business and propaganda interests." (EEIC 1996:9)

Although this cannot be interpreted literally and contradicts several other statements in the strategy document, it nevertheless mirrored views similar to the West's. A view that certain environmental issues are in some way so important that they cannot be compared to other needs of the society.

In similarity with most Western countries, the Polluter Pays Principle (PPP) and the Best Available Technology/Environmental Practice (BAT/BEP) were emphasized as important principles by the Estonian authorities.[116] The PPP states that those responsible for activities that affect the environment in a negative way should bear the costs for the restoration needed. The BAT/BEP principle declares that the most environmentally efficient technical equipment should be used in order to minimize

---

[113] This does not mean, however, that investments in the same issue-area but in different geographical locations are not compared with a cost/benefit approach. This has, in fact, been a very important feature of the e.g. Swedish climate strategy.

[114] Bergström (1994:7), cited in Dabelko (1999:235).

[115] An account of the perception of the environmental aid to Estonia from an Estonian perspective can be found in Tarand (1996).

[116] A brief description of the Estonian experience regarding application of economic instruments for pollution control in the first years of independence can be found in Kallaste (1993).

emissions from point sources (BAT) and that the most environmentally qualified practice should be used regarding diffuse emissions (BEP). The formulation of these principles cannot be interpreted literally. Irrespective of whether it concerns one of the more affluent Western countries or one of the transforming countries, the application of these principles will be related to implementation costs. The interesting point here is that these principles have been made explicit parts of the Estonian environmental strategy, without any explicit qualifications related to socio-economic restrictions. This seems to indicate a strong influence from the environmental regulation approach adopted in the West. It is also the case, however, that these general principles are almost only mentioned in the strategy document. Apart from in the strategy documentation, the focus was usually set on specific emission targets, tax levels etc. This seems to indicate a considerable gap between the general principles and the adopted regulations and guidelines.

Probably the most important factor influencing the Estonian formulation of new legislation concerning environmental protection, the use of natural resources and physical planning, was the need to adapt them to the requirements of the EU (Kallaste et al 1996).[117] The goal set was to have a complete Environmental Code by the year of 2000. To achieve this, it was necessary to reorganize the institutions responsible for environmental issues at both national and regional levels. The institutional system created in the first years of independence consisted of an "institutionally and geographically dispersed management system" (EEIC 1996:40). The goal to have a complete environmental code by the year of 2000 later proved to be overly optimistic.

In Latvia, there had not been any modern environmental strategy formulated before independence. The Cabinet of Ministers formulated the first national environmental strategy plan (NEPP) in April 1995.[118] The first NEPP was to a relatively large extent influenced by current discussions among Western states in general and European countries in particular:

> An environmental strategy is necessary not only for Latvia – it is an essential step in the establishment of a common European environmental system (Ministry of Environmental Protection and Regional Development in the Republic of Latvia (MEPRD) 1995:1).

The conferences of the Ministers of Environmental Protection of Europe held in Dobris (1989) and Lucerne (1993) were both influential for the shaping of the Latvian environmental strategy, as was the OECD initiative Environmental Action Programme for Central and Eastern Europe.[119] In fact, it was explicitly stated in the strategy document that "...the adoption of the NEPP as the environmental strategy

---

[117] A description of the over-all achievements regarding the approximation process vis-à-vis the EU can be found in the European Commission (1998).

[118] The Ministry of Environmental Protection and Regional Development was established in 1993, and replaced the Environmental Protection Committee and the Ministry of Architecture and Building.

[119] All signatories of the Lucerne Declaration assumed responsibility for the creation of a strategy and an action plan. OECD took the initiative for this action program. In October 1997, another ministerial conference was held under the title *Environment for Europe* in Sofia, Bulgaria.

for Latvia is another step for Latvia towards the integration of Latvia into Europe." (MEPRD 1995:1) In order to coordinate the EU approximation strategy, the European Integration Unit was established within the Ministry of Environmental Protection and Regional Development.[120]

The NEPP was the outcome of a series of workshops, to which experts from Sweden and the Netherlands were invited on a consultancy basis.[121] As the preparatory work as well as the actual generation of the Latvian strategy plan were both taking place in close collaboration with political representatives and environmental experts from several Western European states, it seems fair to say that the final content of the plan was also significantly influenced by the current environmental discussions in Europe.

The creation of the first environmental institutions in independent Latvia had an interesting background. Environmental issues were tightly intertwined with the independence movement. Even in the environmental strategy document itself, this was forcefully underlined. Discussing the activities of the Latvian Environmental Protection Club (VAK) it was stated that

> Under the cover of analyzing environmental problems, a number of very important developments were carried out which assisted the re-establishment of national independence ...it is significant that the State Environmental Protection Committee, the first state environmental institution in Latvia's history, was established during this struggle for independence. (MEPRD 1995:4)

The perception of a close relation between the control exercised by Moscow and the lack of environmental concern prior to independence was also clear. When identifying the two groups of important environmental problems to be solved in Latvia – those inherited from the former Soviet Union regime and those connected to the transformation to a market economy - the language referring to the former was very emotional:

> The first group is connected with the grim heritage of the totalitarian period: inefficient, environmentally hostile, resource squandering State-owned industries... the Soviet army left behind a legacy of badly polluted territories (MEPRD 1995:4).

Nevertheless, it is clear that, apart from the instrumental use of environmental issues in the surge for independence, environmental consideration was in fact given increased attention at this time, with large issues like the proposed new subway in Riga and the hydro-power plant close to Daugavpils serving as focal points.

The PPP and BAT/BEP principles were not incorporated in the first Law on Environmental Protection (1991), but were taken into consideration in the 1997 amendment (ECE 1999b). It should be emphasized that these principles, as in Esto-

---

[120] An analysis of the requirements needed for Latvia in each sector in the EU approximation process can be found in ERM (1998).

[121] The content of the environmental plan was exclusively elaborated by Latvian experts, according to the Latvian government (MEPRD 1995:2). Experts from, for example, the Swedish EPA, however, probably influenced the outcome by taking part in the workshops mentioned above. This is not the place, however, to evaluate to what extent.

nia, were more a part of official strategies and proclamations than of actual laws. In many cases, legislation from the Soviet Union was still valid, or had simply been relabeled as national law. The PPP principle implied that those responsible for the environmental damage had to restore the damage or, if this was not possible, compensate the owner (Article 52, Law on Environmental Protection). The application of the PPP principle in the former Soviet Union military sites was rather complicated, since no demands regarding compensation for the environmental damage on these sites were made to Moscow (ECE 1999b: 20). Instead, the Latvian government tried to find private investors to rehabilitate these areas through different kinds of subsidies.

Like the other Baltic countries, Latvia had to harmonize its legislation with the EU in order be able to apply for membership of the union (Linde 1996).[122] Latvia expected to comply with the EU directives in 2010 (MEPRD 1998:6). However, compared to Estonia, which later was accepted as a member in the group of Central European countries that would first be considered as members, the approximation process was implemented at a slower pace.

In Lithuania the first years of independence were likewise traumatic in socio-economic terms. The inflation rate increased rapidly and the gross national product (GNP) shrank abruptly. The economic downturn was even more marked than in Estonia and Latvia, and in 1992, for example, production fell by over 50 percent (ECE 1999a: 3). This caused a sharp decrease in several categories of emission levels. The use of fertilizers and pesticides, for example, was 80 percent lower in 1994 than in 1984 (UNDP 1996b: Chapter 6). However, since the reduction of pollution was generally not caused by implementation of environmental protection measures, the sustainability of these "improvements" was uncertain.

> This "improvement" is, however, very misleading. Because of the economic recession, fewer and fewer resources are being allocated to environmental protection (UNDP 1996b Chapter 6).

In the mid-nineties, the downward trend of the national economy had leveled out, and economic activity began to increase. In 1995 the Lithuanian GDP increased by 3 percent, and by 4.2 percent in 1996 (UNDP 1997 Chapter 2).[123] According to estimates by UNDP, the GDP per capita based on purchasing power parity, was en par with Estonia in 1995, and about 15 percent higher than in Latvia (UNDP 1997).

In 1996, the inflation rate was low, domestic and foreign investments were increasing, the financial situation generally was in order and unemployment was rela-

---

[122] Latvia applied for membership of the EU in October 1995.

[123] The measurement of GDPs is always difficult, especially when comparisons are being made between different countries. In the former Soviet Union countries – as in many developing economies – the comparably large shadow economy makes it even harder to reach adequate estimates of GDP. It has been estimated that the informal economy in Lithuania in the mid-nineties was between 18 and 36 percent of the official GDP (UNDP 1997). This means that (1) the estimate of the GDP – which is based on official figures – probably is too low and (2) the imprecise nature of the GDP figures should be emphasized.

tively low. These were all signs of increased economic stability.[124] The structure of the economy was slowly moving in the direction of mainstream Western economies (i.e. less dependence on industry and agriculture and more on services), although the growth of the service sector was less pronounced than in the other two Baltic States. As in the other Baltic States, the private sector grew. In 1996, more than two thirds of the GNP was produced in the private sector. The corresponding figure in 1992 was approximately one third. In the agricultural sector, private farmers and small landowners produced more than 70 percent of total output in 1992. In 1995, a free trade agreement with the EU was reached, and the following year a similar agreement was reached with the other Baltic States. The comparable positive macroeconomic indicators were, however, paralleled by an increasing socio-economic stratification. Although the wealth of the nation eventually began to increase, this was not the case for large groups of ordinary citizens, who had to face diminishing wages and pensions in real terms, declining levels of social services, and increasing social insecurity.

It is clear that environmental issues to some extent had been crowded out by the main task as perceived by the Lithuanian government: Economic growth in the private sector. In the words of Arvydas Dragunas of the Lithuanian Environmental Protection Ministry:

> The environment was not high on the political agenda, and is not included in the government's list of economic priorities (Dragunas 1995:63).

In 1990, the Department of Environmental Protection was created. This department was not, however, formally part of the government, but was under the direct control of the parliament (MEP 1994b). Four years later, the more comprehensive Ministry of Environmental Protection (MEP) was established as a regular part of the government. The first national Lithuanian environmental protection program was developed in 1992 and the new Law on Environmental Protection prescribed the basic legal principles for the protection of the environment in the same year. The precautionary and the polluter pays principles were envisioned in this framework law. The environmental protection program was cast in fairly general terms, and soon had to be specified and modified in order to comply with the requirements of the EU approximation process (Vebra 1996). A new program was elaborated with assistance from the EU PHARE program in 1995 and accepted by the parliament in September 1996. The 1995 program consisted of three parts: *Strategy motivation*, *Strategy Methodology*, and *Action Programme*. The first volume made an assessment of the present environmental situation in Lithuania, a review of the national economy, a description of the institutional, legal, and economic status regarding issues related to the environment, and a forecast of expected environmental changes. The second part started with an assessment of the present environmental situation,

---

[124] It should be noted, however, that in contrast to the other Baltic States, Lithuania had borrowed quite extensively. In 1997, the domestic debt amounted to almost a quarter of the GNP, and the foreign debt to 20 percent of exports (ECE 1999a: 7)

formulated strategic targets, and defined potential instruments for the achievement of these targets. In the final volume, more precise measures were related to the prioritized environmental problems in order to make implementation efficient. The important issue of how to finance the necessary investments was also discussed.

A comprehensive review of the legislation concerning environmental protection had been given "...highest priority", according to the Lithuanian government (MEP 1996:27). These extensive changes in the environmental legislation ought to be based, according to the 1995 program, on the same (Western) principles contained within, for example, PPP and BAT/BEP as in Estonia and Latvia. One principle, however, that was incorporated into the strategy document and that was not generally found in the environmental regulatory framework of most other Western countries, was the principle of *consistent development*. Although there have been numerous political attempts to direct special attention to less developed regions or economic sectors in most Western countries, they have not been assembled under such a consistency concept. The Lithuanian consistent development principle meant that the uneven maturity of different sectors had to be taken into consideration when formulating sustainable development goals (MEP 1996). The fact that this principle was included in the strategy plan also showed the tension between socio-economic priorities and the requirements necessary to be fulfilled in order to eventually become a member of the EU. A Government Commission for European Integration was established within the Ministry of Foreign Affairs in 1995, and the Legal and European Integration Department was set up in the same year. The approximation process vis-à-vis the EU was the primary mission of both these authorities. In an amendment to the constitution ratified in 1996, it was stipulated that all future legal acts had to comply with EU requirements and directives.

In all three Baltic States, there was a discrepancy between the actual collection of fees, fines, and taxes and the polluter/consumer pays principle. In order to facilitate the implementation of the most urgent investments, it was necessary to rely on other forms of financing. In the first half of the 1990s, most of the financing came from state budget allocations and to some extent from the municipalities, with important additions from foreign sources. The general intent, however, was to increase the level of polluter/user part of financing, with the eventual aim of covering the full costs without external resources.

3. PRIORITIZED ENVIRONMENTAL ISSUE-AREAS

In all three Baltic States, national environmental policy documents were formulated and confirmed by the respective parliaments in the mid-1990s. These documents serve as the main sources for the classification of prioritized environmental issue-areas, and for the description of the national preferences of the Baltic States. These sources do not cover all details concerning different aspects on the formulation of environmental priorities. They do, however, reflect the main characteristics of the political process framing the general environmental priorities in the respective countries. These environmental policy documents are validated against, and supplemented by, documentation from other governmental authorities and external reviewers and scholars. It should be emphasized that the description below refers to the situation *at that time* (1991-96), since this is the relevant data with which to compare contemporaneous Swedish policies. It is further important to note that the purpose is not to describe implemented investments and other types of initiatives, but the composition of the environmental preferences in Estonia, Latvia, and Lithuania respectively. There might be a considerable divergence between preferences and action taken due to a multitude of reasons. Only the preferences, however, are of interest in this study.

As discussed in Chapter two, the national environmental priorities in the Baltic States are assumed to have diverged somewhat from those of Sweden, because of the large difference in GNP. On the one hand, the needs were comparable to those in Sweden a few decades ago, before waste water treatment plants had been constructed in most cities, most harmful point sources had been mitigated etc. On the other hand, the transnational environmental protection discourse put other issue-areas in focus. In order to get into line with the Western environmental agenda, issues such as emissions of carbon dioxide and ozone-depleting substances had to be given due attention, even though it could be argued that at that time these were not the most urgent issue-areas.

It is important to remember that the Baltic States were part of the Soviet Union economy before independence. This meant for example that important parts of the so-called all-union industries were located in the Baltic States, but were as a rule managed by ethnic Russians (Salay et al 1993). Large military facilities were located in this region, as were energy plants (e.g. the Narva power plants in Estonia and the Ignalina nuclear power plant in Lithuania). The Moscow strategy was that these facilities should serve considerably larger regions than the Baltic States. In 1991 Lithuania and Latvia in particular had a significant over-capacity in electricity production, at the same time as considerable volumes of oil and natural gas were being imported from the Soviet Union. After independence, this was a heritage all three states had to handle, albeit in different ways, depending on the nature of the specific facilities. An additional complication in the transition phase of the Baltic countries was the previous large degree of dependence on trade with the Soviet Union. It has been estimated that in 1988, between 85 and 90 percent of all trade in the Baltic States took place within Soviet Union borders (Salay et al. 1993:179). In the energy sector the dependence on Russian exports was especially problematic after independence, due to the significant investments needed to change trade patterns (e.g. in terms of pipe-lines, ports, and other transportation facilities). It has been an important issue for all the Baltic States to decrease their dependence on Russia as a trading partner, both for ideological and pragmatic reasons. The latter have primarily been based on the need to ensure stable flows of goods and services.

A compilation of the environmental issue-areas found most highly prioritized in the official documentation is shown in Table 4 below. As mentioned above, the most important source has been the environmental strategy plans adopted in each of the Baltic States. Each of the categories in Table 4 has been identified as a prioritized area in the official documentation in all three countries. In fact, they were also quite similar to corresponding types of priorities in most Western countries. In some issue-areas, however, subgroups of areas have been of exceptional relevance to a particular country (e.g. nuclear safety in Lithuania, transnational pollution in Latvia, oil shell pits in Estonia).

Each prioritized category is discussed at some length below, and divergences between individual Baltic States are identified. It should be recapitulated that the relative importance of these issue-areas has not been measured. As described above, it is assumed that they have been perceived as approximately equally important by the authorities in the Baltic States.

Table 4. Prioritized environmental issue-areas in the Baltic States according to national authorities.

Nationally prioritized issue-areas in the Baltic States
- Diffuse air pollution
- Point source air pollution
- Diffuse water pollution
- Point source water pollution
- Nuclear hazards
- Solid waste
- Fresh-water resources
- Institutional strengthening and education
- Biodiversity and natural resources
- Landscapes and built-up environment
- Recreation environment

3.1. Diffuse air pollution

Estonia

The major issue regarding air quality in Estonia was, except for the point source environmental disturbances related to the burning of oil shale, emissions from combustion-driven transportation and other emissions of substances affecting climate change and the ozone layer. Generally, the air quality in Estonia was relatively good, due to favorable wind conditions, comparably few large industries (except for the northeastern region), and low population density (EEIC 1996:67). The average wind conditions made Russia the largest recipient of, for example, Estonian sulfur emissions. It has been estimated that almost 40 percent of the discharge was transported to Russian territories, while the Baltic Sea received about 18 percent and Finland 9 percent respectively (ECE 1996:78). According to a survey undertaken in 1994 in Estonia and Latvia, however, air pollution was considered a major issue among large portions of the populations in both countries (Gooch 1995). The emission levels per capita in Estonia were rather high. Approximately twice as much carbon dioxide and two and a half times more sulfur dioxide as the average levels in the West were emitted, primarily due to lack of modern purification equipment and excessive and inefficient energy use.[125] The emission levels were further aggravated by the favored status given to energy based on the burning of oil shale, compared to imported fuels. The price of oil shale and electricity were held at an artificially low level, while the price of, for example, imported oil and gas was not regulated. Since the potential for energy saving was as high as 50 percent on average in the Baltic States, it is somewhat surprising that more attention was not given to the issue of energy saving (Ministry of Industry and Energetics 1993). The energy losses were high in production sectors as well as in, for example, district heating and insulation (Schipper et al 1994; Kristoferson 1993). According to reports from the district heating companies themselves, only about 50 percent of the bills were collected (ECE 1996:74). When the *Riigikogu* (the Estonian parliament) ratified the Convention on Climate Change in May 1994, Estonia committed itself to a reduction of emissions of greenhouse gases. These reductions were later specified to parallel those of the European Union (that is, a 10 percent reduction of CO_2 by the year 2010, and 15 percent by 2015).

The situation in Tallinn was different from most other parts of the country. In particular the carbon oxide concentration was problematic, sometimes exceeding WHO standards by 300 times (EEIC 1996:67).[126] The local air quality in, for example, Kunda and the Narva region was often poor, due to the large cement industry and the burning of oil shale respectively. Goals planned to be reached by the year 2000 included, for example, termination of the sale of ozone depleting substances, stricter restrictions on exhaust gas emissions, and reduction of the use of leaded petrol by 80 percent compared to the 1995 level. By 2010, there should be an 80 percent reduction in the emission of sulfur compounds compared to the 1980 levels. Emissions from the transport sector decreased during the first years of independence, but then began to increase again around 1993-94 (ECE 1996:15). This increase was expected to be augmented further as the economy began to grow faster. In an attempt to decrease emissions from the transportation sector, an import tax of 50 percent was placed on cars older than 13 years. Despite this, since the tax was relatively low, the effect was modest. Taxes on fuels during the first half of the 1990s were almost non-existent, and no extra taxes were levied on leaded fuels.

Latvia

In Latvia, ozone layer depleting substances, climate change gases and regional transportation of sulfur and nitrogen oxides were the highest ranked in the diffuse air pollution category. The activity in this field was relatively limited, though (MEPRD 1996a, 1998). It should be noted, however, that Latvian emissions of SO_X, NO_X, and

---

[125] In 1990, the Estonian energy/GNP ratio was approximately twice that of the United States and three times higher than, for example, that of Denmark. The Latvian and Lithuanian ratios were somewhat lower, but still about twice as high as Denmark's.

[126] The average level of CO was approximately 2000 $\mu g/M^3$ in Tallinn, with peaks of 3500 $\mu g/M^3$ (the WHO ceiling is 10 $\mu g/M^3$ (EEIC 1996:65). The concentration of other exhaust gases has generally been on similar levels as in other Western cities of comparable size.

CO_2 in 1995 were quite low, both in comparison with the other Baltic States and OECD averages.[127] Compared with for example Estonia, Latvian emissions of CO_2 and SO_2 were about one third of Estonian levels. Emissions of NO_x were approximately half the level of Estonia's and Lithuania's, and less than one third of for example Swedish and Danish levels (MEPRD 1996b). Ozone-depleting substances were not produced in Latvia in the first half of the 1990s, but a gradual ban on the use and sale of such substances was envisioned. This issue seems primarily to have been of importance due to the significance of the Montreal protocol, where special attention was given to financing opportunities in the CEE countries. Concerning greenhouse gases, attention was given to the emission of carbon dioxide. The aim was said to be to reduce the level of carbon dioxide to comply with general agreements within the UN Climate Convention. The actual reduction of these emissions, however, depended mostly on modernization and efficiency improvements in energy production and consumption, changes that were probably reasonable even without regard to beneficial effects on the environment. The perceived need to decrease the dependence on external suppliers (i.e. Russia) of energy has been more important for the energy saving initiatives than expected beneficial environmental effects (ECE 1999b: 81). The investments needed in housing insulation and other kinds of restoration were – as in Estonia and Lithuania - very large, but nevertheless attracted comparably low interest. Most interest within the issue-area of transboundary air pollution was attached to regional transboundary pollution caused by sulfurous and nitrous oxides and hazardous substances.[128] Latvia was one of the few net importers of air pollution in Europe. As in the case of carbon dioxide, the primary way to reduce the emission of these gases was to modernize energy and heating sectors, invest in alternative energy sources, and raise consumer tariffs in order to reduce consumption. Relatively few investments were, however, undertaken in this area during the 1991-96 period (MEPRD 1996a, 1998). The number of combustion-driven vehicles decreased considerably in the first half of the 1990s due to the economic slump. In 1995, though, the number of cars increased rapidly, and had risen by approximately 40 percent by the end of the decade (ECE 1999b). The increasing volume of transit traffic since 1993 was considered as problematic from an environmental perspective, even if it was economically valuable.[129] Considerable amounts of freight were transported by railway, but since the locomotives were often old and diesel-driven, this mode of transportation was not always

[127] The emissions of all these gases were lower in Latvia than in either Estonia or Lithuania. Compared to OECD averages, the emission of SOx was a little bit more than a third in Latvia, NOx emissions were approximately one fourth of the OECD's, and CO_2 less than half of OECD's.
[128] Latvia joined the Geneva Convention on Long Range Transboundary Air Pollution in 1994.
[129] In the late 1990s the transportation sector had risen from seven percent of the GDP in 1990 to almost 20 percent.

environmentally sound.[130] It seems clear that the environmental consequences of the increased transit transportation were in general not adequately handled by the mid-1990s, particularly since the measures taken on legislation and regulation were not paralleled by efficient implementation. No reliable data on diffuse air emissions from households and agriculture are available.

Lithuania

The air quality in relation to diffuse emissions in *Lithuania* was, as in most Western countries, mostly affected by transportation practices (MEP 1995a). As in Estonia and Latvia, although car ownership per capita was about half that of an average Western country, the poor vehicle quality and the expected increase in transit truck freight made this issue-area particularly important (Tähtinen 1992). Although the over-all emissions decreased abruptly in the early 1990s, this reduction was almost completely the result of the economic recession. When the economy started to recover in the mid-nineties, emission levels began to increase again, especially from the transportation sector. As early as in 1994, this increase was clearly detectable (ECE 1999a: 77). Pollution from the transportation sector comprised more than two thirds of total air emissions in 1996, which was an increase of approximately three percent since the low-point in 1992 (MEP 1996:9). One of the most highly prioritized items in this area was, consequently, to attempt to halt the increasing trend of pollution from the transportation sector. The major mechanism for achieving this goal was envisioned to be modernization of the vehicle stock and a gradual shift towards fuels less environmentally harmful (e.g. with no lead content). Even if leaded fuels were later officially prohibited (January 1, 1998), there were considerable volumes of low-quality fuel entering Lithuania illegally. The modernization of the vehicle stock was also important from the perspective of the EU approximation process. The estimated cost, however, was quite staggering (estimated to be approximately 5 percent of GDP in 1995), which implied that the implementation of EU standards would take a considerable time.[131] Another prioritized item in this area was the local environmental situation in the larger cities, where emissions from transport vehicles was a considerable local problem (MEP 1995a: Chapter two). Measures were taken to decrease car use, which included economic incentives and organizational management. These measures were, however, not very efficient. Due to inadequate control mechanisms, transport companies did not pay any environmentally related charges (ECE 1999a: 87). The excise duties on fuel did not have any environmental components.

As in the other Baltic States, considerable efficiency gains were attainable from investments and changes of practice in the energy sector and in energy consumption.

---

[130] Latvia is one of the CEE states most extensively endowed with railway networks. These railways were almost completely built during Soviet occupation (i.e. before 1980). Of the total length of approximately 2400 km, however, only about 15 percent was electrified in 1995.

[131] This estimate of the costs was made by the OECD and the Harvard Institute for International Development.

A reduction in energy use also had significant effects on air pollution. It was estimated by the Energy Centre of Vilnius that approximately 50 percent of the energy needed to heat buildings could be saved by efficiency improvements alone. In 1991, the government formulated the program for privatization of the housing sector, and the Law on Housing Privatization was adopted in the same year. Increased incentives for individual saving of energy would have important long-term effects on air emissions in the energy sector. The Housing Renovation/Energy Conservation Fund was set up with an important goal to increase insulation in the house stock. The total cost, however, to provide adequate insulation was as high as US$ 5.5 billion (ECE 1999a: 59). The so-called Bûstas ("Housing") project was introduced in the mid-1990s, with financial support from the World Bank and several bilateral donors. This program was focused on energy savings in the building sector (ECE 1999a). The major problem - how to create private incentives for energy saving - was not solved however. Even though the tariffs increased considerably, thus making it problematic for many consumers to carry these costs, the tariffs were still not linked to the individual's actual use. The main reason for the lack of this link was the fact that few houses had individual meters.

A couple of other joint projects were also started, in most cases in the form of demonstration projects. Since 1993, Lithuania had been a partner in the Swedish Programme for an Environmentally Adapted Energy System in the Baltic Region, whose ambition was to increase energy efficiency, the use of renewable resources etc. In 1996, a national program to make energy use more efficient was adopted, and the Energy Saving Programme Directorate was established in the same year.

Generally speaking, the investments related to reduction of air emissions were relatively modest in Lithuania. It has been stated, "Regulation of pollution from transport and specific pollutants at specific locations should be given much more attention." (Dragunas 1995:67)

Conclusion: The environmental preferences in the diffuse air pollution category were similar in all countries, with transportation practices as a major issue.

3.2. Air pollution from point sources

Emissions from point sources were significantly reduced in all three countries during the first half of the 1990s. The primary reason was the closing or downsizing of many industrial facilities made necessary by the economic recession.

Estonia

The reduction of negative environmental effects from the energy sector has been a highly prioritized area in Estonia. Estonia was, in this period, the largest contributor to air pollution in the Gulf of Finland area (Gooch 1995). In 1993, emissions of sulfur dioxide and carbon dioxide were considerably higher than in most Western states, both in relation to GNP and per capita. Emitters of the most common substances were required to pay charges according to discharged volumes. In 1994, approximately 20 percent of the financial flows into the environmental fund were generated by charges on point source air pollution. Most interest has been focused on the energy sector in general and the burning of oil shale in particular (Liblik & Rätsep 1993). The two largest oil shale based power plants (Estonia and the Baltic) together accounted for approximately 50 percent of particulates and 70 percent of sulfur dioxide emissions (1994) of total point source air pollution in Estonia. The harmful side effects from the energy sector basically consisted of three problem areas: over-consumption, inadequate technological solutions and the use of oil shale for energy production. The excessive consumption patterns were partly a heritage from the price policies pursued by Moscow prior to Estonian independence. The mechanism to curb consumption after independence was to gradually increase tariffs until they covered actual costs. The tariffs rose considerably, but as the apartments and houses in most cases lacked sufficient insulation, the main result was increased consumer costs and insufficient heating. At the same time, the producers were becoming increasingly exposed to competition. A large part of the existing equipment could not be upgraded at a reasonable cost, but had to be replaced completely. The set objective was to reduce dust and ash emissions by 25 percent from the 1995 level by the year 2005, and to reach all levels specified by EU standards in 2010 (EEIC 1996:24).

The burning of oil shale produced grave environmental effects. The two largest power plants based on oil shale provided Estonia with approximately 90 percent of its total electricity requirements, and about two thirds of the energy requirements (Sida 1996:4; Kallaste 1998).[132] Compared to most other CEE countries and like Latvia and Lithuania, the dependence on the use of coal for energy in Estonia has been very low. In 1991, liquid oil accounted for approximately one third of total energy needs, while natural gas accounted for about 13 percent (Salay et al 1993:181). Since most of the oil and natural gas was imported from the Soviet Union, increased trade diversification was perceived to be important. The goal was set by the Estonian government to end all use of fuel oil with high sulfur content (exceeding 2 percent) by 2005. The use of oil shale would, however, continue for the foreseeable future, and further investments targeted at the reduction of air emissions were also expected in the late 1990s (Ministry of the Environment 1998). In order to increase efficiency and decrease emissions, large technological upgrade investments were needed. Import of liquid oil was expected to decline, while the use of peat probably would rise considerably (Tarand 1998).[133] The potential for alternative energy sources was studied, but few projects had been implemented by the early 1990s (Steinrücke 1993; Tomson & Pajumets 1993).

---

[132] According to Estonian authorities, the reliance on oil-shale as fuel for electricity production was as high as 99 percent (EEIC 1996:61)

[133] The burning of peat in considerable quantities was seen as environmentally sustainable by Estonian authorities, because of the corresponding natural growth of the peat reserves. Approximately 22 percent of the Estonian area is covered with wetland (Kallaste 1998).

Latvia

Point source air pollution did not, despite being a prioritized issue-area in Latvia, receive any larger volumes of investments. The primary cause for this was the small number of large emission sources compared to, for example, Estonia. These low emission levels decreased further during the economic slump during the first years of independence. Emission of SO_2 from point sources decreased by about 25 percent between 1991 and 1996 (MEPRD 1996b). Almost all SO_2 emissions were generated in the energy sector. Discharges of NO_X shrank by approximately 40 percent in the same period. Most NO_X emissions originated from boiler houses. The smelter in Liepaja was a particular large emitter of NO_X (MEPRD 1996b). The few initiatives taken in this issue-area were partly a reflection of the fact that in the mid-1990s, the Law on the Protection of the Air in force prior to independence was still valid. Since this law did not require any control mechanisms, air emissions could always be better dispersed rather than decreased by technical devices. According to official statements, point source pollution from energy production was a serious problem, but since it originated to a considerable extent from the burning of oil shale in Estonia, Latvian authorities had no direct control over its sources.

Lithuania

As in Latvia, reduction of air pollution from point sources was not especially high on the Lithuanian environmental agenda, despite being on the list of prioritized environmental issue-areas. In the period 1992-1996, no investments were made targeted at the reduction of the level of air emissions (ECE 1999a: 86). As noted above and in congruence with the other Baltic States, diffuse air pollution from combustion-driven transportation was a large problem. It was stated by the Ministry of Environmental Protection, however, that the sharp drop in emissions from energy production and manufacturing in the early 1990s was almost entirely due to decreased economic activity. The emission of CO_2, SO_x, NO_x, and CH_4 decreased significantly (ECE 1999a). Despite this, some of the larger cities would be classified as "black spots" according to the WHO guidelines on levels of Total Suspended Particles, thus requiring corrective actions. The most significant sources of point source pollution came from the sectors of energy, petrochemicals and metal, and building materials. Most of these facilities were located in the near vicinity of the larger Lithuanian cities. The single most significant polluting facility was the oil refinery in Mažeikiai, which in 1996 emitted about four times as much as the second largest stationary pollution source (ECE 1999a: Table 6.3). It should be noticed, however, that the emissions of CO_2, SO_x, and NO_x (per capita) were still only about half as large as the OECD averages, and considerably lower than those of most other CEE countries (ECE 1999a: Figure 6.1-6.3). As the economy recovered, emission levels were expected to increase, though. It was therefore stated that it was very important to ascertain that the BAT/BEP principles were applied when new investments were implemented. It was estimated that the cost of a significant reduction of air emissions would be quite low. A 40-percentage reduction of the primary pollutants (CO_2, SO_X, NOx, and particulates) could have been achieved at an approximate cost of 0.33 percent of GDP (1995). It was further estimated that energy savings could reduce emission levels by up to 80 percent of the total potential reduction (ECE 1999a: 89). Continuous monitoring of emissions from point sources was almost nonexistent in Lithuania by the mid-1990s. In the typical case, no direct monitoring was performed. Instead, approximate emission levels were calculated from data on input volumes of material and energy, a method producing quite unreliable results.

Conclusion: The major issue in this category was the burning of oil shale in Estonia. Another important concern was whether BAT/BEP would become guiding principles when new investment schemes were to be implemented.

3.3. Diffuse water pollution

Water-related (i.e. point source and diffuse) investments were one of the issue-areas in all three countries attracting the largest investment volumes. In Latvia, for example, 80 percent of the total, environmental expenditures were allocated to this area. The main sources of diffuse water pollution originated from agriculture and transportation. The latter was, however, a secondary effect from air pollution, and has been discussed above.

Estonia

Non point source pollution from agriculture in Estonia was not a significant problem in the first half of the 1990s, due to the sharp drop in economic activity in general and in the use of mineral fertilizers in particular.[134] The most intensive use of fertilizers and pesticides was concentrated on the large state farms. Due to the disappearance of these large entities, the drop in fertilizer and pesticide use was particularly pronounced. In certain areas, however, the quality of the surface water was still deteriorating due in part to the closing of some smaller waste water treatment facilities on former collective farms. In the agricultural sector of Estonia, total production in constant prices decreased by approximately 8 percent between 1994 and 1997.[135] The hazards related to increased pollution when the economy recovered were, however, appreciated by the authorities.

Latvia

The situation in Latvia was quite similar to that of Estonia. In Latvia, the activity in the agricultural sector decreased substantially during the first half of the 1990s. In the period 1986 to 1994, total production in the agricultural sector fell by 50 percent,

[134] In 1995 the annual use of fertilizer per hectare was only one sixth of the volume used in 1988 (EEIC 1996:92)

[135] Source: ESA Statistikaamet, Statistical Office of Estonia.

while the use of mineral fertilizers dropped by 40 percent, organic fertilizers by 30 percent, and the number of cattle by 42 percent (MEPRD 1996a: 45). In 1990, the agricultural sector was the main nitrogen pollution source, emitting 68 percent of total nitrogen discharge into the Latvian rivers (ECE 1999b: 93). Later data have not been available, and it is therefore difficult to know to what extent decreased activity in the agricultural sector has resulted in lower pollution levels of nitrogen. It was apparent, however, that the ecological status of particularly the large lakes around Riga was serious. The eutrophication of Lake Mazais Baltezers, for instance, was still increasing in the mid-1990s (MEPRD 1996b). A target was set in the Latvian Environmental Policy Plan to reduce the discharge of biosubstances from agriculture by 50 percent before 2020 compared to the 1995 level. Another ambition was to decrease the on-going soil degradation, which would require crop rotation schemes and better management of rural landscapes, among other things (ECE 1999b: 145). By the end of 1996, however, investments actually undertaken were modest.[136] Finally, it should be mentioned that compared to many other countries in both Central and Eastern Europe and in the West, the concentration levels of heavy metals in the soil was relatively low in most areas (MEPRD 1996b). This means that potential seepage to nearby watersheds and eventually to the Baltic Sea would be expected to be low (reliable data on such leakage, however, have not been available).

Lithuania

The situation in Lithuania in the first half of the 1990s was similar to that of Latvia. Since nearly half of the Lithuanian territory was used for agricultural purposes, the adverse effects on rivers, lakes, and coastal zones from agricultural run-off were considerable. In 1995 approximately half of the total riverine nitrogen load and about 22 percent of the phosphorous load originated from diffuse sources (ECE 1999a). Even though agricultural production decreased by about 50 percent in the first half of the 1990s and the use of pesticides decreased by as much as 80 percent, the effect in terms of reduced concentrations in the rivers was only slowly becoming detectable. The protection belts erected around lakes and rivers did decrease run-off to some extent, but those belts were in many cases only partly built and not properly maintained.[137]

In all these countries, the need to address the issue of diffuse water pollution in the future was expressed, but few tangible initiatives were taken. It should be noted, that agricultural production has never been expected to recover in the sense that it would return to pre-independence production methods. It was recognized that as the economy became adapted to Western standards, the service sector would increase and the production in the agricultural sector would probably be smaller in terms of

---

[136] In the late 1990s, however, new legislation was drafted in this issue-area, and the Baltic Environmental Agricultural Run-off Project (BEOROP) was initiated in collaboration with Sweden.

[137] No detailed reviews of the efficiency of these protective belts in Lithuania are available. The experiences from similar measures undertaken in southern Sweden in the early 1990s are described in Vought et al (1993). A similar review of the Finnish experience can be found in Uusi-Kämppä & Yläranta (1993).

number of producers. The fact that more modern methods do not necessarily mean less diffuse pollution has not been discussed to any larger extent in the strategy documents.

Conclusion: Emission of nutrients from the agricultural sector was the main issue in this category. Although it was not a significant problem in the first half of the 1990s, because of the decline in production, this was expected to change when the economy in general and the agricultural sector in particular recovered.

3.4. Point source water pollution

Estonia

The condition of the coastal waters in Estonia was seen as a large environmental problem. According to Salay et al (1993:193), "Water pollution [was] perhaps the most acute environmental problem in the Baltic countries." In Estonia, the Gulf of Riga and the Gulf of Finland were severely polluted and eutrophicated due to emissions and run-off from the St. Petersburg region in particular and Estonian cities, and industries. Although the organic pollution load decreased significantly during the economic downturn, the effect on coastal waters quality was quite modest. Apart from the adverse effects on the ecological system of the inland waters, the potential for bathing and other recreational activities was affected. Drainage water from the oil shale pits in the northern part of Estonia was generally discharged into surface waters without any prior treatment, thus causing elevated concentrations of, for example, sulfates and nitrates. Since most of the Estonian population lives close to the Baltic Sea border and major industries typically are located in the vicinity of the larger cities, there was an extensive need for waste water treatment facilities. Due to the economic recession in the first half of the 1990s and investments in sewage systems and waste water treatment plants, the pollution load to the Baltic Sea decreased considerably. However, large additional investments were still needed in facilities treating municipal and industrial waste water as well as in sewage networks, since the waste water prior to independence was generally discharged directly into the sea (EEIC 1996). The praxis of discharging industrial waste water directly to the municipal waste water treatment facilities was a particular problem. A number of projects targeted towards the treatment of municipal waste water were initiated, but few were targeted at a better management of industrial waste water. Most of the initiated projects deployed traditional waste water treatment technologies, but there were also some examples of pilot projects analyzing alternative ways to purify waste water (Ross et al 1993; Priiman 1993).

Significant progress had been achieved concerning policy formulation and redrafting of laws and regulations at the state level by the mid-1990s. A considerable responsibility for the implementation lay with the regional and local authorities. The more precise mandates and responsibilities of these authorities, however, had not been adequately regulated. In combination with deficient funding and a shortage of expertise, this led to a considerable lag in implementation of the decisions taken at state levels. The charge rates, for example, were "...set somewhat arbitrarily by environmental experts without reference either to the economic situation or to the public debate." (ECE 1996:34) In most cases, charges were set too low to have any real influence on consumption or pollution levels. Since acceptable levels of pollution and withdrawals of natural resources always have to be put in relation to the economic as well as the social situation, this tended to decrease the legitimacy of the policy pursued.

Latvia

As mentioned above, Latvian authorities emphasized the transnational component regarding air pollution. The transboundary features of river pollution and the Baltic Sea were likewise strongly emphasized in the Latvian environmental plan. The UN classified three of the large rivers in Latvia as among the largest transboundary watercourses in Europe. The position downstream in a large, transnational river may in some cases be problematical, as was for example experienced by Latvia in November 1990, when an accident occurred in Belarus, causing contamination of the water of the Daugava downstream. The water supply of the cities of Daugavpils and Riga had to be temporary closed due to this accident. Apart from the pollution from other countries carried by these rivers passing through Latvian territory, coastal zones were also often severely affected. Due to wind and current conditions, pollution from Lithuanian sources enters the coastal waters of Latvia to a relatively large degree. Considering all kinds of transboundary water pollution, it has been estimated that only about 50 percent of Latvian emissions to the Baltic Sea originate from Latvian sources (MEPRD 1995:32). Total outlet from Latvian rivers to the Baltic Sea is approximately eight percent of total river discharge (ECE 1999b: 91). The total discharge of waste water decreased by approximately 25 percent during the 1991-1996 period (MEPRD 1996b). Beginning in 1991, the pollution loads from Latvian rivers decreased significantly. This reduction was most pronounced for suspended solids, COD, and BOD(23), while the cutback regarding nitrogen and phosphorus was smaller. It should be noted that initially these reductions did not result in any significant reduction in river concentration levels, and the effect in, for example, the Gulf of Riga was not even detectable (i.e. reduced concentrations were detected, but since natural water discharge was low when the data were collected in 1995, no trends could be validated) (MEPRD 1996b). The most important source of phosphorous emissions was municipal waste water, while the agricultural sector was the largest emitter of nitrogen. In the mid-1990s, investments in waste water treatment plants were either being implemented or planned in many larger cities. In smaller towns, however, few initiatives were taken.[138]

Concerning the larger rivers of Latvia, an additional environmental problem was

---

[138] At this time, though, the so-called 800+ program was launched. This program was targeted at smaller municipalities, and implemented with assistance from the EU. In the late 1990s, however, at least 40 percent of the facilities did not comply with EU requirements (ECE 1999b:99).

the extensive investments in hydroelectric power stations previously implemented by the Soviet authorities, where environmental impact generally was not considered. Also smaller watercourses in many cases had been almost destroyed by ambitions to increase the flows in other rivers, agricultural and forestry output etc. The over-all result was a decline in the number of biotopes and aquatic ecosystems. It should be noted that these kinds of environmental disturbances had almost exclusively national effects, in contrast to the pollution load carried to the Baltic Sea.

Lithuania

In Lithuania, only about one quarter of the total outflow of waste water in 1995 was treated to meet standards in compliance with EU requirements. Reduction of untreated water emissions was consequently high on the Lithuanian environmental agenda, as in the other Baltic States. Approximately one fifth of the total waste water flow in 1995 was not treated at all, while the remaining part was treated, but not to an adequate extent (MEP 1996:8). In 1997, about half of the total flow was treated adequately (ECE 1999a: 94). Although this was an improvement compared with the situation three years earlier, it was primarily due to the decrease of total emissions by almost 20 percent.[139] Most of the sewage water came from the largest cities. About two thirds of the total discharge came from the seven largest cities, where approximately 44 percent of the population lived (ECE 1999a: 93). In the mid-1990s, waste water treatment facilities were being built or upgraded in most of the larger cities. Regarding surface water, the situation was probably most serious in some of the rivers. Roughly speaking, six out of ten were classified as moderately or heavily contaminated in 1995, while most of the lakes were generally affected to a lower extent. The situation of the river Nemunas was quite typical, where the water quality in the uppermost reaches was very good, but deteriorated considerably when passing through larger towns or cities. The level of eutrophication seems to have been almost constant during the years studied. The coastal waters of Lithuania were in some regions gravely polluted by nitrogen and phosphorus. The situation was especially precarious in the Kursiu Lagoon, but other parts were also affected. In the areas adjacent to the outflows of the larger rivers, effects in the form of algae blooming, fish death, and a low sanitary status were clearly visible in the summertime. The discharge of nitrates from, for example, the River Lielupe Basin was estimated to be approximately 20 percent of the total load to the Gulf of Riga (ECE 1999a: 66). The administration of this river basin was especially complex, since this area is divided approximately evenly between Latvia and Lithuania. In 1993, the Lielupe River Basin Commission was established as a joint intergovernmental initiative between Latvia and Lithuania.

A number of prioritized items in this category have been formulated in the Lithuanian strategy plan, but so far most of the resources in this area have been used

---

[139] A reflection of this was the fact that the percentages in the different categories were about the same in 1995 as in 1992 (MEP 1996).

for the upgrading and construction of new facilities for treatment of domestic waste water in the larger cities. These investments have largely been financed by loans, subsidies from the state and from external actors such as providers of foreign grants and international financial institutions.

Conclusion: The most important area in all three countries was the treatment of waste water. In Latvia, the government additionally stressed transnational river pollution as an important issue-area.

3.5. Nuclear hazards

Estonia

The accident in Chernobyl in 1986 placed the issue of nuclear safety at the top of the political agenda. In the documentation of the Estonian environmental strategy, however, nuclear hazards were hardly mentioned at all. There were no nuclear power plants in Estonia. The only radioactive waste generated originated from medical and research facilities, and was on a comparably small scale. Some of the waste material from earlier industrial activity in the Sillamäe metallurgical factory and the former Soviet Union military site in Paldiski, however, hold radioactive contents in varying degrees.

Latvia

In Latvia, it has been noted that there was a small risk with the handling of radioactive substances in laboratories etc, but this was not among the more important environmental problems. Instead, the perceived potential threat came, just as in the case of air and water pollution, from abroad. The largest risks were identified as the nuclear power plants at Ignalina in Lithuania and Sosnovij Bor in the St. Petersburg region in Russia, but the plants in for example Sweden and Finland were also perceived as potentially hazardous to Latvia. The primary means to reduce these kinds of risks have been regional agreements on cooperation, information exchange, and training and education of the inhabitants living near the plants concerning emergency situations in case of accidents. The soil has been contaminated by radioactive substances in some of the former Soviet Union military sites in the Baltic States (e.g. Paldiski Nuclear Submarine Training Centre and Sillamä Metallurgy plant) (VanDeever & Dabelko 1999). Although these facilities have been closed, cleanup initiatives were needed.

Lithuania

The focus concerning nuclear hazards has obviously been put on the Ignalina nuclear power plant in Lithuania, which has some of the largest reactors in the world. Used fuel from the power plant was stored in reservoirs close to the reactor (MEP 1996:11). Solid radioactive wastes were piled up in specially designed storage facilities, while waste water containing radioactive nuclides was stored in a quarry. Environmental effects in the vicinity of the power plant were monitored by plant services, as well as by the Central Environment Research Laboratory and the Institute of Physics. A shortage of advanced instruments, however, meant that the data collected was not fully reliable (MEP 1995a: Chapter 9.1). The environmental effects from the Ignalina plant were not restricted to the risk of an accident and constant low-level emission of radioactive substances. The large use of cooling water resulted in "thermal pollution". The temperature of the surrounding Druksiu Lake was elevated by approximately three degrees, which tended to increase eutrophication (MEP 1996:20). It should be noted, that other than environmental considerations have been important in relation to the Ignalina plant. As most of its staff was Russian, it served as a focal point for the independence movement and was later considered by some groups to be a potential security risk to the Lithuanian Republic.

There was a quite intense debate among Western countries concerning the safety and future of the Ignalina power plant.[140] It was contested whether the recommendation should be to close the plant entirely or to update its safety standards. Although it did not seem reasonable to close the plant immediately, since Lithuania depended greatly on its production of electricity, the question was left open by the Lithuanian government.[141] It was simply concluded, that: "In case a decision is taken to shut down the nuclear reactors, a program for this purpose should be elaborated containing safety measures to be employed during the closing-down period." (MEP 1996:20) The situation regarding the future operation of the Ignalina power plant was unclear, primarily because of the difficulties in predicting the level of investment support from abroad. In 1994, it was assumed by the Lithuanian Deputy Minister of Energy, Mr. Saulius Kutas, that "...within five or six years...we will have to select the most advanced, reliable and safe reactor world-wide, the type which will replace the current one." (Kutas 1994) In late 1999, the Lithuanian government stated that one of the two reactors would be closed down in the year 2005. This decision was confirmed by the parliament in the spring of the year 2000.[142] The EU had put considerable pressure on the Lithuanian government, stating that the closing of the Ignalina power plant was a prerequisite to becoming a member of the union. Lithuanian authorities, mapping the potential effects from a complete close-down of the two reactors, worked out different scenarios. All these scenarios showed that a shutdown of the Ignalina power plant would increase air emissions considerably. According to the Lithuanian Energy Institute, it would be preferable to continue to keep Ignalina open for its complete lifetime, but with increased measures taken to safeguard against accidents and other environmental hazards (ECE 1999a: 83).[143] The future of the power plant was, as late as the spring of 2000, not completely

---

[140] See for example Connolly & List (1996).

[141] Approximately 90 percent of Lithuanian requirement for electricity was supplied by Ignalina, making Lithuania the country most dependent on nuclear energy in the world (ECE 1999a:155).

[142] This decision was taken with 67 votes for, and 25 against.

[143] A scenario of the potential effects from a serious accident at the Ignalina Power Plant is presented in MEP (1995a: Chapter 9.2).

clear, given that reasonable alternatives were not available.

Conclusion: The future of the Ignalina power plant in Lithuania was the major item in the nuclear hazards category not only for Lithuania, but also for Estonia and Latvia.

3.6. Solid waste

Management of solid waste was a significant environmental problem in all three Baltic States. Attempts were made to find a joint management solution, in order to reap economies of scale benefits. However, no such initiative had been implemented by the end of the 1990s.

Estonia

In Estonia, the management of waste at civil as well as military sites has historically often been inadequate. Particularly in former Soviet military sites, the most immediate aim was to investigate the extent and type of waste deposits. Former Soviet military sites covered almost two percent of Estonian territory at approximately 1,500 locations. Discharged aircraft fuel posed a particularly severe problem, because of its high toxicity. The rehabilitation cost for the former Soviet military sites was staggering, and by the mid-1990s few initiatives had been implemented. The allocated expenditure for the 1992-1995 period corresponded to about 0.03 percent of expected total costs (ECE 1996:42). Similar needs for improved management practices were also relevant regarding, for example, industrial sites and mines. The hazardous waste generated by the oil shale industry posed one of the most severe problems in this issue-area. Almost all hazardous waste generated in Estonia originated from the use of oil shale. Seepage from the highly alkaline ash tips affected the fresh-water supply of the Narva region, as well as the Bay of Finland (Roose 1991). The mining of oil shale was often undertaken in open pits, and in the early 1990s almost 7.5 percent of Estonian territory was directly affected by the mining of oil shale (Salay et al 1993). Waste material from the mining, as well as from energy production, were as a rule piled up in large stacks, making seepage of heavy metals and other toxic substances substantial. The size of these stacks was considerable. The landfill area of Kiviter, for instance, covered about 250 ha and was about 100 meters high. Although no reliable data existed, it was clear that the volume of generated waste was considerable. Since the charges were based on generated volumes, there was a strong incentive to under-report on generated waste volumes. It was not until the mid-1990s that the first waste incineration facilities were constructed. Before this, all waste was just dumped (ECE 1996).

The liability claims have been a particular problem, since it was often hard to find responsible individuals or authorities. The general aim was, according to the national environmental strategy document, to significantly reduce the generation of waste. The municipal waste was dumped in landfill sites, without any attempts to sort different kinds of waste. Industrial waste was in most cases also dumped at the municipal landfills. The number of landfills was, according to an inventory undertaken in 1995 almost 450, and in many cases quite close to a municipality. Of these landfills, more than half were unauthorized. It would not be possible to use most of these landfill sites even under a new, environmentally adapted framework due to, for example, their geographical location close to cities. The target was set to decrease the number of sites to 150 by the year 2010, and the long-term ambition was to reduce them to 20 (EEIC 1996:96). It should be noted, however, that as in Latvia and Lithuania, the problem was related to management rather than to generated volumes. The volume of municipal waste generated per capita was below the OECD average (336 kg/capita compared to 383 kg).[144] According to the management of waste policy adopted between 1991 and 1993, the responsibility was shared between several authorities (i.e. the ministries of environment, economics and agriculture, the counties, and the municipalities). This dispersion of responsibility made it hard to enforce actual implementation of laws, regulations, and guidelines.

Latvia

The management of solid waste was a highly prioritized item also in the national, environmental strategy of Latvia. The resources spent on this issue-area were, however, quite modest. In the 1993-1996 period, the expenditure on waste management was between 0.9 and 4.5 percent of the total state allocation to environmental issues (ECE 1999b: Table 3.6). Although the generation of waste in 1995 was significantly lower in Latvia than in most Western countries, waste management was at best rudimentary.[145] It has been estimated that the investments necessary to meet EU requirements would be in the range of € 230-550 million (ECE 1999b: 129). It was further expected, that an adaptation to Western consumerism would probably increase the waste volumes considerably.

In 1996, 550 landfills were in operation and at least 160 additional sites had been located and closed for further use (MEPRD 1998). Even in 1997, none of those in operation were managed in accordance with appropriate environmental procedures (ECE 1999b: 124). A number of illegal waste dumps had been identified, but their content was in most cases not known. Since no national plan on waste management had been implemented until 1998, almost every municipality created its own dump-site. In the typical case, no environmental considerations were taken while managing the landfill, and with only one exception (the *Demene* site in Daugavpils), hazardous waste was mixed with all other kinds of waste. In the law on municipal waste that was being drafted in the late 1990s, only hazardous waste was taken into account. One of the gravest hazardous environmental problems was the approximately 2,000 tons of discarded pesticides, which were kept at about 420 different sites. Most of them were in poor condition (MEPRD 1998). Seepage from these landfills, irrespective of whether they were currently in use or not, threatened to contaminate

[144] Source: ECE (1996).

[145] In 1995, the generation of waste from households and services in the cities was between 150 and 300 kg/capita and year. Corresponding figures for Western Europe were 300-450 kg/capita and year. Source: Ministry of Environmental Protection and Regional Development and Eurostat.

groundwater and soils (Schrøder & Kjærgård 1993). In 1997 a national management strategy was elaborated and accepted by the Cabinet in 1998. The management plan is now binding for all waste producers, waste management organizations and appropriate political bodies at relevant levels. The waste management strategy focused on three main targets: reduction of environmental impact from illegal dumping, reduction of seepage from official dumpsites, and decrease of waste production and increase of recycling. In general, most households in the countryside did not have any waste collection. In the towns, most apartments were connected to such a service, while most independent houses were not. A key issue regarding waste management in general was undoubtedly to make those responsible for the generation of waste pay for its management. Regarding non-hazardous waste, this issue had not been properly addressed by the late 1990s.

The handling of hazardous waste was given particular attention, not only due to its potential effects on the environment and human health, but also as it was identified as an important item by the EU in the approximation process of Latvia. Until 1995, no management plan concerning hazardous waste existed. Therefore, the national hazardous waste management system was elaborated in cooperation with representatives from the EU and the Danish government. In 1997, a Hazardous Waste Project Implementation Unit was founded. The first task for this unit was to compile a list of locations of hazardous waste dumps, the composition of the waste, storage etc. It was found that the main categories were discarded pesticides, PCB, and waste from the health care sector.[146] Some other important types of hazardous waste were combustion waste and slag, sludge from waste water treatment, and oil slime.[147] The management of hazardous waste was, however, still not fully in line with EU directives, since its definition was not unambiguous (ECE 1999b: 121). As in all the Baltic States, there was often incongruence between legal requirements and actual implementation. It was stated in the 1993 Law on Hazardous Waste that generation, transportation and deposition of hazardous waste had to be reported. Actual reporting, however, did not start until 1997.

Lithuania

The situation in Lithuania was quite similar to that of Estonia and Latvia. Although the production of domestic waste per capita here was also considerably lower than in most Western countries, it was perceived as important to make projected increases as low as possible. The generation of waste decreased significantly during the first years of independence, but as this was due to the economic downturn rather than investments or changes in practice, it was expected that the volumes would increase

---

[146] The Implementation Unit evaluated 423 different sites, and found a total of approximately 1,400 tons of old pesticides. More than half of this volume was found in the Rezekne county in the east of Latvia (MERPD 1997b).

[147] In 1997, a register of contaminated sites previously managed by the Soviet Union military was set up. A total of 255 sites were identified, with varying degrees of contamination. The old military sites had not been cleaned up by the end of the 1990s, however, due to lack of state financial resources.

when the economy started to grow. Sorting and treating different types of refuse in adequate ways could have reduced the volumes of waste considerably (MEP 1995a: Chapter 8.1). The investment in waste management in the period 1991-1995 was almost non-existent. In 1996, investments were about 10 percent of the total environmental expenditure, while the figures for the water and air sectors were 66 and 23 percent, respectively.[148] In the late 1990s there were still 900 registered landfills in Lithuania, and probably a significant number of smaller, unregistered sites in the countryside. In most cases both types of sites were located in the vicinity of towns or villages. In 1992, guidelines for the selection of dumpsites were established by the Ministry of the Environment and Ministry of Health. The main problem remained however, namely the issue of how to finance the remediation of the many sites. Before independence, the general method was simply to collect all domestic waste, and dump it somewhere outside the city. In order to achieve a more comprehensive treatment of domestic waste, it was seen as imperative not to leave the full responsibility to the municipalities. In fact, "... it is irresponsible to leave the problems related to domestic and other kinds of non-hazardous waste management to the municipalities alone", according to the environmental strategy plan (MEP 1996:20). The tendency towards the end of the period under study, however, was to decentralize authority to the regional and municipal level, and in the Law on Environmental Protection it was stated that the municipalities were responsible for the management of household waste.[149] In the late 1990s, however, the charges collected for waste management still did not cover the costs (i.e. landfill constructions and maintenance). The major initiatives planned to be launched in this area were expected to be focused on targets such as the formulation and adaptation of regulation and management of dumps, increased recycling, and the renaturalization of closed dumpsites.

Regarding other non-radioactive hazardous wastes, it was decided by the parliament in 1991 that individual plants should store the hazardous waste themselves until a national strategy on management issues had been implemented.[150] The Hazardous Waste Management Programme was adopted in 1993. The centralized approach in this program, however, was later replaced by a more decentralized one, which was supposed to be considerably cheaper. The investments needed for the remediation of the many dumpsites were staggering. It was estimated that the rehabilitation of the former Soviet Union military sites alone would amount to about US$ 1 billion (ECE 1999a: 126). The deposits of old or prohibited pesticides constituted one of the gravest environmental problems in Lithuania (MEP 1995a: Chapter 4). The general trend in the second half of the 1990s however, seemed to be a de-

---

[148] Source: ECE (1999a).

[149] In 1997, the municipalities had the managerial responsibility for the waste, water, and heating sectors (ECE 1999).

[150] A program had already been set up in 1982, where the Palemonas Ceramic Factory was supposed to take care of certain kinds of hazardous wastes. Only a small fraction, however, of the hazardous waste arrived at Palemonas during the first half of the 1990s due to many enterprises' inability to cover the costs for waste treatment.

crease in the generation of hazardous waste. It was estimated that the total of stored pesticides amounted to about 4,000 ton. More than half of this amount was characterized as "... prohibited, old or unclassified." (MEP 1996:11) No less than 26 sites caught fire during the period 1989-95.[151] Another important category of hazardous waste in Lithuania was oil-containing substances. It was estimated that this type of waste comprised almost 60 percent of the total generation of hazardous waste (ECE 1999a: 120). A particular kind of waste-related problem was acute in the Aukštrakiai dumpsite outside Šiauliai, one of the most heavily polluted areas in Lithuania. Large volumes of toxic, liquid waste generated in the tanning industry had been put in several ponds. An accident in this area could result in major outflows of chromium, sulfides, and organic matter into the surrounding small rivers. These pollutants would then quite easily be transported to the nearby Baltic Sea coastline. Hazardous wastes from hospitals were generally dumped together with ordinary domestic waste. The environmental problems connected with these kinds of hazardous waste were not solved, despite the fact that dumping on municipal dumpsites was made illegal. Instead, the result was that hazardous waste was piled up in storage facilities of varying quality. It was estimated that approximately 70 percent (1996) of the hazardous waste was accumulated on-site (ECE 1999a: 39). It was also possible that a relatively large part of it had been dumped either in the dumpsites or elsewhere (MEP 1996).

Conclusion: A major issue-area in all three countries was the management of the numerous waste sites but reliable information on their content was not available.

3.7. Fresh-water resources

Estonia

In the early nineties, approximately two thirds of the drinking water consumption in Estonia came from the groundwater supplies. In general, the supply of clean water was relatively satisfactory, except in certain areas (e.g. in the capital Tallinn, where water quality was sometimes rather poor, resulting, example, with an outburst of Hepatitis-A in 1993).[152] In a study undertaken in 1993 by Birk et al, it was concluded that "... existing water purification technology does not guarantee the epidemiological safety of drinking water." (Birk et al 1993:273)[153] Emissions from the oil

---

[151] The Government Programme for the Management of Obsolete Pesticides adopted in 1997 seemed, however, to be successful. By the end of the decade, most piles of old pesticides had been identified and repacked. It was expected that a significant proportion (approximately two thirds) of these pesticides could later be put into use, while the remaining portion would have to be incinerated.

[152] The Health Service, under the Ministry of Social Affairs, was the authority responsible for monitoring drinking water quality.

[153] See also Leesment & Raud (1993).

shale industry in the northeast resulted in polluted groundwater.[154] Since the northern part of Estonia is the least endowed with groundwater, the situation was problematic at several locations. Apart from affecting the water quality, mining operations also lowered the groundwater table, making extraction more costly. Close to the former Soviet Union military sites, oil had been found in the groundwater, and pollution from farming made the water impossible to use in some areas. The quality of the groundwater may deteriorate further due to pollution not yet fully investigated. The water supply system was generally of very poor quality. Leakage of 30-35 percent was not uncommon, and in the rural areas, only about 60 percent of the population was connected to the public network. The most immediate measures to take were to reduce consumption, by, for example, imposing tariffs reflecting actual cost and to upgrade distributional networks. In the mid-1990s, it was estimated that necessary investments for the provision of good-quality drinking water would be approximately 1 billion SEK (ECE 1996:32). During the first years of independence, water consumption fell dramatically by about 35 percent (ECE 1996:24). Most of the fresh-water was used for cooling and processing in the industrial sector. In 1993, the total use of drinking-water per citizen was more than twice that of the average citizen in the West, despite being only about five percent of the total fresh-water consumption. The aim was set to reduce network leakage by 50 percent before the year 2000. The pollution of surface waters has decreased considerably since Estonia gained independence. This was, as in Latvia and Lithuania, primarily due to the sharp decrease in economic activity in the industrial as well as in the agricultural sector.

Latvia

Fresh-water resources were in general quite good in Latvia. Only about 1.7 percent of the available resources were used in the mid-1990s (ECE 1999b). The supplies were large enough to provide people with drinking water, provided that the supplies were used reasonably. Despite these favorable conditions, excessive use of groundwater resources created depressed water levels in certain areas (e.g. in Riga and Liepaja). In the larger cities approximately 95 percent of the population was connected to a central network system, while the corresponding percentage for the inhabitants in the middle-sized towns was about 50 percent. In the countryside, however, most people used dug wells, with an average depth of about 2-6 meters. From a national perspective, groundwater and surface water was used for drinking purposes to approximately the same extent. Despite the fact that the potential supply should meet demand, two major categories of problems had to be handled by the contemporaneous water regime, namely pollution of raw water and malfunctioning distribution facilities.[155] Although the groundwater reservoirs had not been seriously

---

[154] The Geological Survey Centre was the main monitoring agency of groundwater quality.

[155] The Ministry of Welfare was responsible for drinking water quality, while the State Geological Service was the agency accountable for groundwater issues. No program had, however, been implemented targeted at the quality of drinking water in countryside wells.

affected in most places, considerable concentrations of harmful substances had been observed in some areas, for example around the Jurmala municipal waste disposal site (*Kudra)*, the Riga waste disposal site (*Getlini*), and in the vicinity of the Tukums oil storage site (MEPRD 1996b). An indication of the importance attached to the preservation of high-quality drinking water was the comparably high levels of charges for water pollutants implemented in the mid 1990s.[156] It should be noted, however, that groundwater pollution from the agricultural sector was in general lower than in most Western countries.

The distributive network system was generally of inferior quality, and its management has often been inadequate. This resulted in for example an increased risk of bacteriological contamination. The dug wells common in the countryside were often of inadequate quality. The shallow depth of the wells made them unreliable sources for drinking water, since chemicals and bacteria could easily contaminate them. The quality of the water in the dug wells was investigated in the period 1992-1994, and it was found that 17 percent of the wells were polluted with nitrates and nitrites (ECE 1999b: 95).

Lithuania

The protection of fresh-water resources has been made a priority item also in the Lithuanian environmental strategy. A large proportion of the Lithuanian watercourses is a direct result of human activity. Only about 15 percent are natural rivers, while the remaining 85 percent are either regulated parts of rivers or land reclamation canals (MEP 1996). Producers of energy have heavily dominated the water usage. Of the approximately 4.5 billion cubic meters of water extracted annually in Lithuania, almost 92 percent was used in the energy sector primarily for cooling and cleaning purposes. Of the remaining water, somewhat less than 5 percent was used by the domestic sector. The withdrawal of groundwater has not been a general problem, although in the vicinity of some of the major cities withdrawal rates adversely affected the groundwater level. The reliance on groundwater was however relatively low in Lithuania. Of all water used, less than 7 percent was extracted from the pool of groundwater. Regarding its quality, approximately one third of the Lithuanian area has been estimated not to have usable fresh water. Since most households in the countryside rely on water from dug wells, the pollution of groundwater might have serious health effects. It was estimated in 1996 that more than half of the wells were not adequate sources for drinking water from a health perspective. Drinking water provided by central sources was generally of higher quality, but it was still estimated that about 40 percent of the actual water supply did not meet hygienic standards (primarily because of a high iron content).[157] The Regional Public Health Centres regularly monitored the quality of the central supplies of drinking water, while sam-

---

[156] Comparing the periods 1993-1995 and 1996-1997, the charges were between three and four times higher in the later period. Source: ECE (1999b). These charges were not indexed, however, but had to be changed by the parliament.

[157] Source: Data from the National Centre of Hygiene and from the Ministry of Health (1994).

ples from dug wells were taken occasionally by the same authority. In the recession of the first half of the 1990s, water usage dropped by more than a fifth. The intensity of use (i.e. the use of water in relation to available resources), however, was still quite high, about 50 percent higher than the OECD Europe average and about twice as high as the OECD total average (OECD 1997).

One of the most urgent measures to implement in this sector, according to the environmental strategy plan, was to induce a more rational use of fresh-water resources. The most obvious first step would be to install water meters on all intakes (MEP 1996). This would make it possible to implement efficient taxation of the natural resource of fresh-water. Another target of considerable importance was to upgrade the fresh-water network, which would make the losses of water during transportation smaller.

Conclusion: With the exception in certain areas, there was no real lack of fresh-water in any of the countries. Potential leakage from waste sites, however, could become a problem. The distribution networks were very often in poor shape in all three countries.

3.8. Institutional strengthening and education

In all three countries, the most important issue in this category was the formulation of laws and regulations compatible with EU standards. This was a highly prioritized issue in all the countries. Although the EU approximation process had just begun, it was quite clear that an even more problematic issue was to create a reasonably efficient and reliable public administration. The adopted legislation on environmental taxes and consumer charges was quite similar in all countries, and clearly adapted to Western standards. The collection of fees and fines, however, in many cases was inefficient if not completely absent. Due to insufficient administrative capacity, judicial enforcement powers, and in many cases unclear ownership, the implementation of the rather strict legislation was only partial. In many cases the charges were set too low to have a real effect, making it beneficial for companies not to implement environmental investments. It should be noted, though, that in the second half of the 1990s this problem was addressed and charges and fines of several kinds were increased. They were not, however, increased sufficiently to cover all external costs.[158] As an example, it was concluded by the Economic Commission for Europe that in Latvia "... the pollution charges are relatively low, so that polluters do not fully incorporate environmental externalities in their costing procedures." (ECE 1999b: 51) One of the consequences of the inability to make the polluters pay was that the state in effect subsidized the enterprises. Another consequence of

[158] One example of a relatively successful legislation and implementation was the Latvian Law on Natural Resources Taxes. This was, however, adopted in 1995, but it was not until the revised version was in force in early 1997 that it had any real impact. Between 1996 and 1998 tax revenues from natural resources increased from 0.5 to 1.5 percent of the state budget (ECE 1999b:38).

probably even greater importance was that the polluters' incentives to decrease emissions were lower than would have been optimal if environmental externalities had been fully internalized, thus making the investment level in more environmentally-friendly equipment and management sub-optimal.

Both local and state authorities were responsible for the return flow of these resources to the environmental sector. In Estonia, the taxes were funneled to the Estonian Environmental Fund, established in 1983 (Ministry of the Environment 1992). This fund was reorganized as an extra-budgetary fund in 1990, and reorganized once again according to the Act on the Environmental Fund in 1994. In Latvia, the Latvian Environmental Investment Fund (LEIF) was not introduced until 1998.[159] In Lithuania, the State Nature Protection Fund (SNPF) was established in order to collect fines from those using natural resources above the permitted level. Regular taxes from the use of natural resources, however, went to the state budget. In addition to the SNPF, the Environmental Fund for Investments (EFI) was established to allocate state budget means to environmental investments.[160] In all the Baltic States, the intention was to channel environmental taxes and charges back to the environmental sectors at the local, regional, and state levels (Ministry of the Environment, Estonia 1992, Environmental Protection Committee, Latvia 1992, Environmental Protection Department, Lithuania 1992). In all three countries, an administrative problem in the first half of the 1990s was the large proportion of revenues channeled to state authorities, rather than to regional and municipal levels. In Estonia, for example, approximately 80 percent of the revenues from environmental taxation were appropriated by the government. It was expected, however, that about 70 percent of these revenues were to be channeled to local budgets in the future.

Regarding regulation on the use of natural resources and permission to pollute, all the Baltic States tried to find an efficient mix of economic incentives and administrative control mechanisms. The policies adopted were, however, usually more precisely elaborated in theoretical rather than in practical implementation. In Lithuania several business activities, e.g. prospecting and extracting natural resources, construction of state roads, export of timber etc, which might have adverse environmental side effects, required a permit. The permits to exploit natural resources and the pollution of the environment were based on regulations of the Ministry of Environmental Protection (MEP) and on recommendations by HELCOM. If a company had such a permit, environmental side-effects had to be self-monitored and reports

[159] The Public Investment Programme (PIP) was, however, already established in 1995. Since then, environmental investments at the state level were channeled through the PIP (EC 1999b: 47). In the late 1990s, the target was set to allocate one percent of the state budget means through the PIP. It should be noted, however, that the total investments decreased abruptly in the period 1994-1997, despite having increased steadily in the 1991-94 period. In 1998, though, the level of environmental investment in Latvia increased considerably (ECE 1999b:50).

[160] In addition to these funds, there were 55 local funds at the municipal level in Lithuania, and a number of sectorial funds (e.g. Forest Fund, Road Fund, Energy Efficiency Fund, Health Fund etc). As in Latvia, the over-all allocation priority scheme was elaborated in the Public Investment Programme (PIP).

issued to the MEP. In terms of actual implementation, however, the efficiency of this regulation was poor, since it was quite rare for companies to achieve a permit. In Vilnius, for example, it was concluded that only about one percent of the enterprises had applied for and received a permit (ECE 1999a: 29).

As the share of the financing of environmental investments from foreign sources (grants and loans) increased during the first half of the 1990s, it became increasingly urgent to establish an efficient coordination of these monetary flows.[161] Not least the donors wanted to make the effects of their contributions as large as possible, and thus required coordination with other providers of assistance. This was achieved in somewhat varying ways, either through the establishment of a coordination unit or by increased contacts between the authorities involved. In Lithuania, for example, a project management group was set in up 1994 by the MEP in order to coordinate assistance from a multitude of sources. The establishment of this management group was an assisted project in itself, partly financed by EU PHARE funding.

Estonia

The promotion of public, environmental awareness was stated as a priority in all three countries. In Estonia, the need not only for public information, but also public participation in different decision processes was emphasized. By the mid-1990s, however, information on for example nature conservation was rather limited (ECE 1996:90). A higher degree of environmental awareness was supposed to be attained through education and training programs. For example, "…a general environmental protection class will be included in the curricula of universities, high schools, and vocational schools." (EEIC 1996:52) The state was furthermore expected to take an active part in changing consumption patterns into more environmentally sustainable patterns, by using mass media resources, economic tools, and administrative instruments (EEIC 1996:20). Estonia was participating in the preparation of the so-called Guidelines on Access to Environmental Information and Public Participation in Environmental Decision-making under the auspices of the Economic Commission for Europe (ECE 1996). Environmental NGOs played an important role in the independence process of the Baltic States. However, the role of the environmental NGOs became less important by the mid-1990s, partly because public interest decreased and partly because former NGO experts often were employed in the administration (ECE 1996). The two most influential Estonian organizations, the Estonian Nature Conservation Society and the Green Movement were both particularly active on issues concerning nature conservation.

Regarding the ratification of environmental conventions, Estonia had by the mid-1990s not signed the Geneva Convention on Long-Range Transboundary Air Pollution, nor the Convention on Environmental Impact Assessment in a Transboundary

[161] It should be noted that even if the level of external assistance rose during the period, the funding from enterprises increased considerably more rapidly. In Lithuania, in 1996, 12 percent of total environmental expenditures came from donors and IFIs, while enterprises contributed with approximately 60 percent (ECE 1999a: 32).

Context nor the Vienna Convention for the Protection of the Ozone Layer.[162] Among the signed and ratified conventions were the Basle Convention on the Control of Transboundary Movements of Hazardous Wastes and their Disposal (1992), the Washington Convention on International Trade in Endangered Species of Wild Flora and Fauna (CITES, 1992), the Bern Convention on the Conservation of European Wildlife and Natural Habitats (1992), the Ramsar Convention (1992), the Framework Convention on Climate Change (1994), and the Convention on Biological Diversity (1994). It should further be mentioned that Estonia adopted the UNCED Act on Sustainable Development in 1995.[163]

Latvia

As has been mentioned briefly above, environmental issues played a prominent role in the Latvian independence movement. The environmental awareness among ordinary citizens was regarded as comparatively high. In a survey conducted in 1995 by UNDP, few respondents gave "don't know" answers although a majority (55 %) were of the opinion that their "...information level about environmental problems [was] low." (UNDP 1996a)[164] It was also a fact that in the monthly polls on the popularity of ministers, the Minister for Environmental Protection was always close to the top (ECE 1999b: 53). The interpretation of such results is, however, somewhat problematic, since it is not clear to what extent this reflected environmental concern and to what extent it rather mirrored the legitimacy of the independence movement of the early 1990s. In the mid-1990s there were approximately 60 environmental NGOs active in Latvia. Comparably few of them exerted influence by lobbying for increased environmental protection etc.[165] Some of the NGOs did, however, partly shape the contemporaneous environmental policy, since the most qualified representatives often became enrolled by the government, regional authorities, or by the municipalities.

The creation of the Latvian National Environmental Policy Plan (NEPP) was not characterized by active public participation. The NGOs had no official representation, and business interests were hardly represented at all (ECE 1999b). Another important group with little influence on the formulation of the environmental policy in Latvia was the large Russian minority. Partly depending on the environmentally concerned actors' close relation to the independence movement and partly due to the weak representation in political bodies in general, the Russian minority was almost completely absent in the environmental discourse of independent

---

[162] Regarding the Ozone Convention, Estonia did, however, not produce any chlorofluorocarbons or halons, and the consumption of these substances was limited (ECE 1996).

[163] A description of the early Estonian policy on sustainable development can be found in Oja (1993).

[164] See also Gooch (1995) and Loftsson et al (forthcoming).

[165] This was partly due to the lack of a centralized agency for the environmental NGOs and partly due to the fact that most of them were more focused on environmental education and rehabilitation than on lobbying.

Latvia (ECE 1999b: Chapter 4).[166] This lack of involvement of important citizen groups and interests was problematic, and tended to undermine credibility as well as stifling the potential to implement specific parts of the NEPP.

Environmental education played a relatively modest role in undergraduate education. The topic had no formal status, and was not given an explicit prioritized status by the Latvian government (ECE 199b: 57). The main problem was the lack of textbooks and inadequately paid and insufficiently trained teachers. At the university level, the most important institution in environmental research and education was the Centre for Environmental Science and Management (CECAMS), which was part of the University of Riga. It should be noted, that CECAMS was originally an environmental NGO, but was later transformed into a multidisciplinary research facility. Regarding the publication of official data on environmental issues, the Latvian Environmental Data Centre was relatively successful. Although the data were not always reliable (they were, for example, sometimes based on companies' emission permits rather than on actual monitoring), they were nevertheless important from a public information perspective. Data from the Environmental Data Centre were synthesized and discussed by the Environmental Consultancy and Monitoring Centre, and publicized in state-of-the-environment reports. This Centre also played an important role as provider of environmental information at public request. In similarity with CECAMS, the Environmental Consultancy and Monitoring Centre is also a former NGO that has been transformed into a public agency.

Finally, regarding international environmental cooperation, Latvia signed the Rio Declaration in 1992. The Ministry of Environmental Protection and Regional Development was responsible for the implementation of the Agenda 21. Latvia was also one of the eleven members of Agenda 21 for the Baltic Sea Region, established in 1996, where Latvia, together with Germany, was responsible for the transport sector. In 1992, the Basle Convention (hazardous waste) was ratified by Latvia, and the United Nations Framework Convention on Climate Change was signed in the same year, and ratified by the parliament in 1995. In the same year, the Regulation on the Accession to the Vienna Convention and Montreal Protocol was ratified by the Latvian parliament.[167] Also in 1995, the Convention on Biological Diversity was ratified, and the National Biodiversity Action Plan was formulated in cooperation with the IBRD.

Lithuania

The Lithuanian government issued similar statements on public information, as the Latvian authorities did. It was for example envisioned that relevant parts of the recommendations of the UN ECE Convention on Environmental Information

---

[166] According to surveys undertaken by UNDP in the mid 1990s, there was no indication that concern about the local environment systematically varied between different ethnic groups, which indicates that the difference in activism was more a reflection of political accession (UNDP 1997).

[167] Later the same year, however, all the three Baltic States notified the UNEP that the implementation of the protocol would be postponed until the year 2000 (ECE 199b:69).

Dissemination should be implemented in an attempt to increase public awareness and environmental education. Special attention was given to the education of institution officials as well as teachers and students. Courses on cleaner production and waste minimization were given at the Kaunas University of Technology in 1994 as a joint project together with Lund University in Sweden. The Lithuanian Agriculture University and the Vilnius Gediminas Technical University were also starting courses on clean technology. Lithuanian universities were also partners in the Baltic University Programme, where for example sustainable industrial production was on the curriculum. The availability of expertise on environmental management was scarce in Lithuania in the period under study. In addition to the lack of expertise, interdepartmental cooperation was less developed than in most Western countries (ECE 1999a). This was an impediment to an efficient strategy concerning integrated management of natural resources. According to the Lithuanian environmental strategy, one aim was for example "... to introduce ecological education into the curricula of all levels of educational institutions." (MEP 1996:35) In the Law on Environmental Protection (1992), it was stipulated that the government was obliged to make environmental information (environmental yearbooks, state of the environment reports and similar kinds of documents) available to the public. According to the same law, the citizens had the right to take part in discussions on environmental projects, request state ecological examinations, carry out similar initiatives themselves etc. The actual implementation of these goals was, however, quite modest. In Vilnius, for example, although the municipal authorities collected data on emission levels, these were not made available to the public, except for some brief information on the Internet. The information through mass media and other public channels was also quite modest (ECE 1999a). Among the more prominent environmental NGOs, the Green Movement, the Lithuanian Fund for Nature, and the Lithuanian Environmental Protection Society should be mentioned.[168]

Concerning the adoption of international conventions and international agreements, Lithuania signed (1992) and ratified (1995) the United Nations Framework Convention on Climate Change. Since 1994, Lithuania had also been a part of the Vienna Convention for the Protection of the Ozone Layer, and ratified the Montreal, Copenhagen, and London amendments. In the late 1990s, it did not, however, fulfill the Montreal requirements (ECE 1999a: 69). It has also ratified the Convention of Conservation of European Wildlife and Natural Habitats (1996), the Convention on Biological Diversity (1995), and the Ramsar Convention (1993). Lithuania signed the Long-range Transboundary Air Pollution Convention in 1993, but had not ratified it by 1999. The Convention on the Protection and Use of Transboundary Watercourses and International Lakes was also signed but not ratified (1999). The Convention on Environmental Impact Assessment in Transboundary Contexts has not yet been submitted to the parliament (1999). It should also be mentioned that

[168] The Regional Environmental Center for Central and Eastern Europe (REC) has also played an important role, but its activity in Lithuania was, until the late 1990s, limited.

Lithuania has been a member of IAEA since 1993.[169] At the regional level, the participation in the Environment for Europe process was important with regard to adaptation of Western principles for sustainable development, even if lack of resources was a considerable hindrance to their actual implementation. Lithuania furthermore signed the 1992 Helsinki Convention, but this was not ratified until 1997.[170] It has not (1999), however, ratified the related International Convention for the Prevention of Pollution from Ships (MARPOL) or the Convention on Civil Liability for Oil Pollution Damage. In 1995, an agreement was reached between the three Baltic States on environmental cooperation. Lithuania was, finally, a member of the Baltic Environment Forum (BEF), which is a joint initiative between the Baltic States and the EU. Lithuania was also taking part in the Baltic Agenda 21 process.

Conclusion: The most important issue was undoubtedly the approximation process vis-à-vis the EU, but also environmental education was given considerable attention. An over-all shortage of expertise, financial resources, and institutional capability resulted in a considerable gap between conventions, laws, regulations, and procedures and their implementation.

3.9. Biodiversity and natural resources

Estonia

The protection of biodiversity and natural resources has been set as a priority area by the Estonian authorities.[171] Several species, e.g. bear, wolf, and lynx, which are very rare in other parts of Europe, show comparably strong populations in Estonia (EEIS 1992). Approximately 12 percent of the land had been classified as protected areas in the early 1990s. The Estonian nature conservation initiatives were especially targeted at the protection of wetlands. Under the National Programme for Implementation of the Convention on Wetlands with International Importance measures had, by the end of 1996, been taken in the Matsalu reserve. Further initiatives were planned in Käina Bay, Alam-Pedja Nature Reserve, Saaremaa National Park, and in the Endla, Nigula, Muraka, and Lallatu reserves (EEIS 1999). Due to earlier restrictions on the use of coastal areas, large parts of these areas were less affected by human activity than coastal zones of many other European countries. The most important threats to the biodiversity and traditional landscapes were perceived as the general transformation and diminution of the agricultural sector and

---

[169] The international regulations ratified by Lithuania concerning issues related to nuclear material were incorporated in the Law on Nuclear Energy in 1996.

[170] Lithuania is, together with Finland, the Lead Country in the sustainable forestry sector in the HELCOM Baltic Sea Joint Comprehensive Environmental Programme (JCP).

[171] A compilation of the most important aspects of the Estonian National Biodiversity Action Plan as perceived in 1995 by the Estonian Ministry of the Environment, can be found in Külvik & Tambets (1995).

the increasing economic activity in the coastal zones. It was stated in the strategy plan that the "...negative trends are enhanced by the persisting social attitude of the mid-1990s which underestimated the need for preservation of biological diversity." (EEIC 1996:64) Another potential threat was oil spills from sea vessels, an issue that has occupied much interest from HELCOM ever since the formulation of the 1974 Convention. In common with the other Baltic countries, Estonia was highly dependent on Russian oil and natural gas. It was perceived as important to diversify the procurement of energy supplies as a step towards national sovereignty. This meant, together with increased trade with the West, that considerably larger volumes of oil, for example, would be transported to and from ports in the Baltic States.[172] By 1993, this increased transportation had already caused environmental disturbances. According to P. Ojava, the Estonian Marine Inspection, "The increasing number of incoming tankers (every seventh incoming vessel entering the port [of Muga] in 1993) has caused a drastic increase in the number of accidents and thus the increase of sea pollution incidents." (Ojava 1994)[173] Considering the fish stocks and seal populations in this area (the coastline of southern Estonia and northern Latvia) and, of probably greater importance, the migratory paths of rare waterfowl, this would potentially be a serious threat to Estonian and Latvian biodiversity. According to A. Kuresoo et al, "Even small oil spills can cause serious damage to these populations..." (A. Kuresoo 1994)

Estonia is relatively sparsely endowed with natural resources. Apart from oil shale, the most important were probably the large reserves of peat and wood, which could be used for energy production. Forests covered approximately half of the Estonian territory in the mid-1990s. Between 1990 and 1995, the use of wood and peat for energy production increased from three to eight percent of the total energy supply. Although the use of peat can be regarded as sustainable, provided that the extraction is lower than the natural growth of the reserves, there were several environmentally disturbing side-effects from burning peat briquettes and other kinds of peat inputs. Depending on the efficiency of the burning process, emissions of metals, NO_X, and particles varied considerably (Kallaste 1998).

Latvia

Reduction of threats against biodiversity was also seen in Latvia as an important part of the over-all environmental strategy, and a relatively large number of initiatives were taken early on to safeguard against further deterioration (MEPRD 1998). As

[172] Due to their geographical location, the ports of Estonia could be increasingly interesting for the exportation of Russian oil. According to O. Tammemäe, Head of the Department of EIA and Normatives in the Ministry of the Environment, Estonia, Russian producers (e.g. from Komi and West-Siberia) had in 1994 already shown such an interest. (Tammemäe 1994).

[173] The construction of the port of Muga illustrates a common phenomenon in the Baltic States. The port facilities had already been planned by Moscow in the 1980s. The project met fierce opposition, however, from the Estonians, since it was seen as a Russian project that would not benefit the Estonians in any way. After independence, the potential gains from a port in Muga became more clearly visible.

briefly mentioned above, the policy formulated by Moscow resulted in areas with very high economic activity, while other regions were left almost untouched. Therefore, and as in Estonia, some species and biotopes that were very rare or extinct in most other parts of northern Europe still existed in Latvia. According to the strategy plan: "This fact increases international responsibility for Latvia for the preservation of biodiversity from the continental perspective." (MEPRD 1998:48) The most influential factor concerning the protection of sensitive habitats in Latvia was, as in the other Baltic States, seen as the general development of the practices in agriculture and forestry. Latvia signed the 1992 Convention on Biological Diversity. It had not, however, adopted any of the EU regulations in this area (1999). Nor had it established competent authorities regarding the handling of genetically modified organisms, as required by the EU (ECE 1999b: 22).

Unlike Estonia and Lithuania, Latvia has few domestic energy resources and was therefore dependent on supplies from other countries. The main geological resources were peat, which was available in relative abundance, construction materials (e.g. limestone, gypsum, clay etc), and high-quality mineral water reservoirs (MEPRD 1996b). Although it was appreciated in the early 1990s that geological oil reserves might exist, it was estimated that no more than approximately ten percent of the Latvian energy requirement could be supplied by domestic sources (Kudrenickis et al 1994). It was also appreciated, however, that exploitation of the potential oil resources could seriously threaten important biotopes and thereby Latvian biodiversity. As in Estonia, an increased dependence on oil transported over the sea and an enlarged transit trade threatened biodiversity in the coastal zones. A former Soviet military base, for example, had seriously contaminated the Liepaja seaport. A potential rehabilitation of the port was now considered. The Liepaja region is sensitive from a biodiversity perspective. The coastal waters are important spawning grounds for herring and some other species, the coastline is attractive from the recreational perspective, and important pathways for migratory birds pass over the Liepaja region. According to Eistere and Junkurs (1994), "...crude oil terminals...in Liepaja will have an adverse impact on the unique coastal natural complex and may cause unrecoverable changes to it." The pollution of the Gulf of Riga was particularly critical, since it was the most productive fishing ground in the Baltic Sea (ECE 1999b: 111).[174] It was at the same time severely polluted, and regarded as one of the most seriously affected areas in the Baltic Sea. Regarding other biotopes, probably the two most important were the wetland systems and the meadows. The former were seen as important from a biodiversity perspective, but were threatened mainly by the extraction of peat. The extensive meadows (approximately 12 percent of the territory) were rapidly transformed as the traditional cultivation practices were abandoned.

In 1996, about seven percent of the Latvian territory was protected. The selection of territories to protect had been an ongoing process for the previous 50 years. The

[174] Fish and fish products have been one of Latvia's most important export commodities. Although production decreased significantly in the first half of the 1990s, it started to increase again around 1995-96.

reasons behind their protection, however, have been many and various, and not always because of environmental considerations. Therefore, an extensive revision of the most important areas to protect was of significant relevance (MEPRD 1998:52). It is clear, however, that enforced property rights from the era before Soviet occupation would make the protection of valuable natural and cultural areas considerably more difficult (Gooch 1995).

Lithuania

Lithuania has ratified the Ramsar Convention on Wetlands of International Importance Especially as Waterfowl Habitat, and its amendments.[175] In the mid-1990s five sites had been notified on the Ramsar list: Čepkeliai, Kamanos, Viešvilė, Žuvintas, and the Nemunas Delta. Traditionally, large areas in the Lithuanian countryside have been managed as meadows, thus producing substantial biodiversity. Recently, however, these valuable sites have to a considerable extent been lost due to diminishing agricultural activity.[176] This trend towards overgrown meadows began, however, before Lithuanian independence. In the period 1956-1980, the meadow area decreased from almost 20 percent to a little over 6 percent of the total territory. Biodiversity in urban areas was, despite the effects from industrial activities and high population density, comparably large. In the mid-1990s, biodiversity and conservation issues were becoming increasingly addressed (MEP 1995b). The Biodiversity and Conservation Strategy and Action Plan was adopted in January 1998 (ECE 1999a). Biodiversity issues have not, however, been systematically monitored hitherto, even though flora and fauna have been monitored from other than biodiversity perspectives. The authority responsible for environmental monitoring in Lithuania was the Joint Research Centre at the Ministry of the Environment. Regarding environmental NGOs, they played quite a prominent role in this issue-area, as most of them were especially interested in biodiversity and nature protection issues (ECE 1999a). It has been stated by several NGOs, however, that information from government sources was not sufficient. An attempt was made to also incorporate ecological considerations in the State Tourist Strategy, as developed by the Lithuanian Tourism Board. A practical implementation of this approach would be, for example, to concentrate on so-called eco-tourism. The Law on Tourism (1998) did not, however, regulate these kinds of activities. Considering that it could be expected that most of the tourists would visit protected areas, a conflict between tourist and commercial interests and nature protection was expected.

In all countries, the potential long-term threat of global warming has been appreciated. The projected effects were, however, not only negative from the perspective of the Baltic States. The agricultural and forestry sectors would probably generally benefit from an increased average temperature, as would the tourist sector (Kallaste 1998).

---

[175] Contemporaneous prevalence and diversity of bird species is described in MEP (1995a: Chapter 6.2.1).

[176] For a general description of the situation regarding prevalence and tendencies in the number of mammals of different species, see MEP (1995a: Chapter 6.1).

Conclusion: Protection of the relatively undisturbed areas in the respective countries was seen as an essential issue in this category. The on-going transformation of traditional agricultural practices and life-styles was perceived as a major threat to biodiversity.

3.10. Landscapes and built-up environments

Apartments, work places, and other buildings were in all three countries often not built to satisfactorily reduce the level of noise. In some regions, in-door radon levels posed a serious threat to human health, and in other places the constructions did not meet requirements necessary to protect other aspects of human health.

Estonia

In the early 1990s, there were 4 national parks, 5 state reserves, and 50 other types of protected areas.[177] Among the 50 protected areas, 14 were classified as landscape reserves (ECE 1996:49). The areas with status as strictly protected were only 1 percent of all protected locales, but this percentage was scheduled to increase to 5 percent by the year 2010. The tendency in the mid-1990s was to increase local and regional responsibility for the protection of nature. Due to a shortage of administrative capacity, expertise, and smooth communication channels with researchers and other experts, however, implementation lagged behind. Despite having ratified for example the Convention of Biological Diversity and the Ramsar Convention, there were for instance no general guidelines distributed to regional and municipal levels by the mid-1990s. In a similar way, the implementation of the 1994 Act on Protected Natural Objects and the 1995 Act on Sustainable Development was limited in its scope. The open oil shale pits also posed an environmental problem from a natural landscape perspective (Toomik, Samlan & Kaljuvee 1993). Traditional agricultural landscapes were seriously affected over large areas. It was estimated that the mining of oil shale had destroyed at least 11,000 hectares. Some of these areas had been recultivated, but the original landscape had nevertheless often been lost.

Latvia

In Latvia, the protection of landscapes and built-up environments was seen as a priority. In 1996, more than two thirds of the Latvian population lived in towns and approximately one third in the capital, Riga. The result of this urbanization was not only a general impoverishment of the cultural landscape, but also polluted watercourses, abandoned gravel pits, inadequate land reclamation and abandoned territories formerly occupied by the Soviet army (MEPRD 1996a). In the Law on Environmental Protection, it was established that if an area was regarded as having

---

[177] The primary responsibility for the implementation of the nature protection policies was within the Environmental and Nature Protection Inspectorate and the State Marine Inspectorate.

exceptional cultural or aesthetic values, the consent of the landowner was not required for its protection. The law did, however, stipulate possibilities for tax exemptions and compensation to the landowner. Regarding Latvian built-up environments, the towns of Sigulda, Cesus, and Valmiera in the northern part of Latvia all have a rich history, worth protecting. In the eastern part of Latvia, the town of Kuldiga exhibited a similar cultural heritage, which might be threatened if protective measures were not implemented. A mix of traditional agriculture and small industries characterized several areas in the countryside. The initiatives taken to protect valuable areas in the countryside were, however, very few (ECE 1999b). Regarding natural landscapes, the riverbanks of the Gauja and Amata were especially valuable.[178] Another landscape worth mentioning are the river valleys of Venta and Abava, situated close to the town of Kuldiga in the western part of Latvia. Although it was perceived that short–term measures were necessary in order to protect valuable landscapes, general economic development was also seen as the decisive factor by Latvian authorities. An over-all "balancing of the economic activity" would be necessary in order to ensure a sustainable development (MEPRD 1996a: 53).[179]

Lithuania

The protection of landscapes and built-up environments was only vaguely referred to as a priority issue-area in the documentation of the Lithuanian environmental strategy. It is nevertheless clear, that it was seen as important to preserve, for example, valuable cultural environments, and that this might contradict the urge for economic development. A comparably large proportion of Lithuanian landscapes had been protected. Approximately 11 percent of the territory was protected by the mid-1990s, including five national parks, 30 regional parks, 6 strictly protected reserves, and 290 general reserves of different merit. Of the general reserves, no less than 61 had been established due to their landscape characteristics and 10 for their cultural values (MEP 1995a: Chapter 7). Certain areas were exempted from the privatization process, thus making it possible to establish recreational areas. A serious threat to the landscape and other environmental assets was posed by potential seepage from the numerous dumpsites. Toxic substances from such leakages could easily be transported by nearby rivers and by the groundwater and thereby affect extensive areas. Apart from ensuring the protection of natural and biological assets, one national and one regional park had been given special status as exhibiting historical and cultural values deserving of special protection. There were several outstanding old towns and historic monuments in Lithuania. In 1995 Vilnius was included on the UNESCO list of World Heritage Sites. Due to insufficient resources, however, renovation of buildings was often restricted to facades and the ground floors where businesses were located. In the late 1990s, approximately 7 percent of Lithuanian territory was covered by marshes and bogs, which significantly shaped the natural

[178] In the late 1990s, a biosphere reserve was established in this area (ECE 1999b).
[179] A compilation of sustainable development projects implemented in Latvia before 1997 can be found in MEPRD (1997a).

landscape in these areas. Increased efficiency in agricultural methods (e.g. draining of bogs), however, threatened this type of landscape (Balčiūnaitė 1993). About 15 percent of the agricultural landscape was further classified as severely eroded. An additional five percent was subjected to moderate erosion. When damage was done to the landscape, compensation could be demanded according to contemporaneous laws. This compensation was funneled to the State Nature Protection Fund. Depending on where the damage occurred, the level of compensation differed. For example strictly protected reserves or national parks, the recompense rose tenfold (ECE 1999a: 27), while in other types of less strictly protected areas the compensation was multiplied by a factor of three or five.[180]

The regulation of territorial planning was of great importance for the potential to protect valuable landscapes and built-up environments (MEP 1995a: Chapter 7). This regulation was, however, adopted rather late. The most important laws with a bearing on spatial planning were the Law on Territorial Planning (1995), the Law on Construction (1996), and the Law on Environmental Impact Assessment (1996). At about the same time, the Environmental Impact Assessment Division was established within the MEP, with the mission to secure the protection of valuable landscapes, protected areas, and ecosystems. It should be noted, that, even though the contemporaneous work on spatial planning had commenced somewhat late, Lithuania did have a prior history of territorial planning. The spatial planning not only for Lithuania but also Latvia, Estonia and the Kaliningrad region had been located in Lithuania during the Soviet Union's occupation.[181]

Conclusion: All three countries expressed similar interest in the preservation of the landscape and valuable cultural buildings. The most serious threats were perceived to be the lack of financial resources and the general modernization of the economy.

3.11. Recreation environment

This final category of prioritized environmental areas in the Baltic States has not been extensively discussed in the documentation of the Estonian and Latvian environmental strategies. This, however, does not mean that recreational environments were not seen as important issue-areas, but rather that they were primarily discussed in other contexts. As in most modern Western countries, recreational environments were seen as important from an environmental viewpoint and as an integral part of a welfare society.

---

[180] The level of compensation was regulated in resolution 198 (1995).

[181] The Urban Development Institute (Vilnius), the Architecture and Construction Institute (Kaunas), and the Industrial Planning Development Institute (Kaunas) were the Soviet Union authorities responsible for spatial planning in the Baltic region.

Estonia

Pollution of Estonian coastal waters, caused primarily by inadequate waste water treatment facilities, affected the potential to use, for example, beaches for recreational purposes. Occasionally, the authorities closed several beaches because of poor water quality (ECE 1996:29). The many lakes in Estonia covering about two percent of the territory were valuable assets from a recreational perspective. Many of them, however, were adversely affected by primarily industrial discharges of pollutants. Several fish species have decreased in numbers, or disappeared completely. The potential for spawning has been seriously affected in many rivers, and several previously important salmon rivers were now devoid of any salmon. In the early 1990s, almost a quarter of the Estonian territory was classified as protected zones of different merit. A large part (2/3) of this area, however, was not protected according to IUCN standards (ECE 1996). Most of these locales were protected because of their valuable coastal vegetation or peat bogs. The Matsalu State Nature Reserve, which is on the Ramsar list, was protected because of its wetlands, which served as important bird sanctuaries. The main threat to the nature reserves and other recreational areas was inadequate control over the economic development in the early 1990s. Land reform processes and big changes in ownership rights resulted in major transformations, which meant that satisfactory management of protected areas could not be fully maintained. Some areas, the Lahemaa National Park for example, with important recreational values, were becoming increasingly affected by commercial interests, e.g. tourism projects, that endangered important nature value. Another example was the threat posed by uncontrolled draining of mires and meadows. It was the explicit ambition of many counties to develop tourism strategies. A shortage of financial resources and a lack of administrative capacity and expertise, however, often resulted in a ruined landscape and adversely affected biotopes when these strategies were implemented.

Latvia

In Latvia, several areas where traditional agriculture and small-scale manufacturing was practiced were valuable from a recreational viewpoint, as were the riverbanks and valleys of Gauja, Amata, Venta, and Abava. The Gauja national park was also important as a winter sport resort. The other of the two Latvian national parks is Kemeri. Somewhat less than seven percent of the territory in the mid-1990s was classified as a protected area. Some of these areas were privately owned, and rules and regulations were imposed on the owners. However, compensation to these landowners was quite insignificant, which made the incentives to protect these areas weak.[182] The vast forests of Latvia are not only important from an economic perspective, but also as a valuable recreational source. In 1996, about 40 percent of the Latvian territory was covered by forests.[183] Unlike many regions in Western Europe,

---

[182] The Law on Specially Protected Nature Areas was adopted in 1993. This law, however, did not specify provisions for the compensation of landowners.

[183] Source: State Forestry Service of Latvia.

the Latvian forests were typically composed of many species and had varied characteristics. In the mid-1990s, not enough financial resources were made available to the municipalities for the management of specially protected forests. The result was that the municipalities were not particularly interested in getting forest areas classified as protected.

The many lakes of Latvia were valuable for their recreational value as well as for tourism.[184] The ecosystems of the lakes and their surroundings were generally very sensitive and it was therefore seen as important to carefully plan investment projects. The importance of protecting such areas had, however, to be weighed against the need for industrial investments, housing, infrastructure and similar issues, which were all part of the general economic development. Along the shores of the Baltic coast and the Bay of Riga, a 3.000-meter wide protective corridor was created, where the landscape was not to be affected. Although this protective belt was primarily established to preserve biodiversity, it was also important from a recreational point of view. Relatively few tourism investments were implemented along the Baltic coastline, partly due to previous restrictions caused by the large military installations. Although some private sector initiatives were taken, the activity in the tourism sector was quite modest. Despite the establishment of the protective belt along the coastline, the sand dunes etc were continuously being adversely affected because of the poor enforcement capability of the official authorities. In the second half of the 1990s, the tendency seemed to be to relax restrictions regarding the coastal corridors (ECE 1999b: 116).

Territorial planning undoubtedly was a key to the balancing of different, and sometimes conflicting, interests. According to a review of the Latvian environmental performance by a UN agency, however, "Territorial planning does not seem to be a top priority in Latvia's transition policies" (ECE 1999b: 33). Since natural areas important for recreation but not sufficiently unique to be covered by laws on special habitats were not legally cared for, they were not adequately managed or protected. It should further be noted, that the expenditure on national parks and reserves declined in the 1991-1996 period. In 1991, 1.4 percent of the total expenditure was allocated to this area, while the average percentage in 1992-1996 was only about 0.4 (ECE 1999b: 50). A similar trend was occurring in the area of land recultivation, where the corresponding percentages were 0.9 and 0.2, respectively.

Lithuania

It was estimated that about one third of the total Lithuanian territory could be classified as recreational areas. The spatial structure of Vilnius's landscape, for example, was almost identical to that of the 16th century (ECE 1999a). Several areas of valuable landscapes in Vilnius and its surroundings were, however, threatened both by the recession – with its accompanying decreased public expenditures - and economic

---

[184] The Ministry of Environment and Regional Development was responsible for state policies on tourism in Latvia.

and social transformation. Laws on spatial planning were adopted quite late (1995-96), which made it harder to protect valuable recreational areas in this period. More recently, the government has proclaimed ambitions to establish programs for the protection and development of green areas in the cities, and to restrict traffic in the old town centers. Up till now, however, few initiatives in this direction have been put into effect. The most frequently visited locales, the Baltic Sea coast, the lake areas in the east and the suburban areas were receiving considerably larger numbers of visitors than other recreational sites. The frequent use of these locales resulted in increased pollution and transformation of the environment, since intensified recreational activities often brought in their wake construction of leisure housing and other facilities. Due to the large volumes of discharged, untreated waste water, the beaches of, for example, Klaipėda, Giruliai, and Palanga were not suitable for bathing due to high levels of microbial contamination. Therefore, bathing was in most cases prohibited.[185] The many regional parks attracted large numbers of visitors. In the mid-1990s, it was estimated that these parks had about 11 million visitors annually. The exact number of visitors was not known, however, since surveys had not been undertaken on a systematic basis. The most valuable areas had to some extent been protected by legal jurisdiction, but: "Land reform has been a hindrance to the protection of most valuable recreational resources, sometimes totally obstructing it." (MEP 1996:13)

Conclusion: The importance of recreational environments has been expressed in similar terms in the three Baltic countries. Despite this, only small amounts of public expenditure were used for increased protection and enhanced management of recreational environments.

4. CONCLUSIONS

Apparently, there was a considerable degree of homogeneity in environmental preferences among the Baltic countries. They all expressed similar environmental strategies as in most Western countries. The potential to finance the required investments was, however, considerably more restricted.

The approximation process towards eventual membership in the EU was a "meta-strategy", at a level above the more specific issue-areas, as were the attempts to introduce modern incentive-based regulations, such as the PPP principle, consumer charges covering the actual costs of public services, and environmental taxes. Similar strategies regarding the funding of environmental investment were furthermore adopted. Environmental funds were established for the collection of revenues from pollution taxes. At a later stage, increased resources were allocated at regional and local levels. All countries had considerable difficulties in actually collecting consumer charges and emission taxes. Environmental issues related to point source

---

[185] The responsible authority for the quality standards of recreational waters was the Centre of Ecological Medicine of the Institute of Hygiene.

emissions to air and coastal waters received the largest resources in all the countries. Issues concerning nuclear hazards attracted significant resources in Lithuania, due to the Ignalina power plant. Considerably smaller amounts were targeted at, for example, fresh-water resources, waste management, biodiversity, landscapes, and environment recreation assets. A common institutional problem was the divergence in administrative capacity and expertise between, on the one hand, state authorities and on the other hand regional and local bodies.

A few issues diverged from the general pattern of homogeneity of preferences. The burning of oil shale in Estonia for the production of energy was a particular environmental problem for Estonia, although neighboring countries were also affected. The nuclear power plant in Ignalina was primarily a dilemma for Lithuania, although a serious accident could have affected the whole region. Latvia was particularly exposed to transboundary pollution, due to its location downstream of several large rivers.

Before the preferences of the Baltic States described in this chapter are compared to actually undertaken joint projects, a description of how the Swedish support program was administered will now be given. An account of what issue-areas Sweden perceived as the most important ones will also be presented.

CHAPTER 6

THE CONTENT AND ADMINISTRATION OF SWEDISH ASSISTANCE TO THE BALTIC STATES

1. DECENTRALIZED DECISION-MAKING

In this chapter, the primary ambition is to describe the administration and content of the Swedish support. This is relevant for the validity of the model, since these issues constitute the most important link between the interests of the key decision-makers and the implementation of the chosen strategy vis-à-vis the Baltic States. The main purpose is not, however, to detect bureaucratic distortions in the governmental implementation, but to show that the CEE assistance primarily was not implemented by the main foreign aid agency, Sida, but rather by agencies specialized in transfer of technology, financing, and advanced international cooperation. This shows the notion described earlier that the environmental assistance was perceived by the governments as well as by the other parties in the parliament as something different from traditional foreign aid. The choice of implementing agencies thus indicates something about the strategies chosen by the parliament. The most important reason why the focus is not on the potential influence of the bureaucracy on the actual implementation phase is that the prime interest is not in the outcome, in terms of impact in the Baltic States. Rather, the focus is set on the preferred policies of Sweden.

From the early start of the Swedish assistance program, the responsibility for its implementation was divided between several political and administrative bodies. At the political level, the main responsibility lay within the Ministry of Foreign Affairs. The Secretariat for Cooperation with Central and Eastern Europe (ÖSEK) was established in 1989 for the specific purpose of administrating Swedish CEE support. Originally, it was placed under the Trade Department (Avdelningen för Handel) at the Ministry of Foreign Affairs, but three years later it was made an autonomous unit directly under the Ministry of Foreign Affairs (Stockholm Environment Institute 1994:6).

Initially, the support to the CEE countries developed quickly and at that time no comprehensive strategy of its implementation had been formulated. Due to the great pace of the changes taking place in the former Soviet Union by the turn of the decade, the approach by the government could be characterized as reactive. A need for supportive action was clearly felt, although it was somewhat unclear what kind of assistance would be appropriate. The government adopted a view, however, that the assistance should be quite broad in range. The needs in the countries in transformation were of many kinds, and the best way to support them was by paying attention to many of these needs. The result was that important decisions affecting the character of the early support program were taken by several different political bodies. Apart from the Ministry of Foreign Affairs and the Cabinet, also the Ministry of Environment and Natural Resources, the Ministry of Commerce, and the Ministry of Transport and Communication soon became involved in the CEE support program.

Another important factor when interpreting the composition of the Swedish support program was the decentralized politically-bureaucratic decision-making. Swedish ministries are relatively small compared with most other countries, while the implementing agencies are quite large and have a considerable degree of autonomy. While the political bodies of the ministries set the goals, general priorities, and funding schemes, the administrative agencies are not only responsible for implementation, but also have a significant discretionary powers in project approval and elaboration. In the typical CEE assistance case, the authorities in the recipient country took the first initiative.[186] When selecting a potential joint project, domestic priorities in the recipient country were the point of departure. In order to increase the likelihood of approval from the donor country, however, an assessment of that country's preferences also had to be made. This meant that donor country considerations were already to some extent taken into account in the recipient country's selection process. High priority projects would not be proposed as suitable for foreign assistance, unless a positive response by the donor was a reasonable probability. Because of this, the project proposals by the potential recipient country did not reflect the national environmental priorities in any straightforward sense, but were rather a mix between domestic priorities and probability of approval.

In this chapter, the projects supported by Sweden will be briefly described. These will be placed in their context within particular programs, as administered by different agencies etc. In the next chapter, the projects will be categorized into different groups, according to the type of environmental enhancement they were targeted to achieve. If not otherwise explicitly stated, the total amount of the Swedish contributions refers to the period 1991-1996. 1996 has in other words been taken as the cut-off point, even in those cases where future Swedish assistance was envisioned.

2. THE SIGNIFICANCE OF THE BALTIC SEA ENVIRONMENTAL ACTION PROGRAMME

The activities of the Helsinki Commission (HELCOM) and the Helsinki Convention have often been seen as somewhat of a cause behind the increased level of international, environmental cooperation in the Baltic Sea region.[187] This is a reasonable premise in many cases, depending on what aim the analysis has been designed to achieve. It should be emphasized, however, that HELCOM is a very small organiza-

---

[186] According to some sources, it was not uncommon in the first years of the Swedish CEE program, however, that the original initiative came from a donor country's consulting corporation. This has been described in the case of Swedish environmental support to Poland. See for example Radetzki (1994).

[187] See for example Hjorth (1992; 1995) and Pedersen (1994).

tion, and has no implementation capacity of its own. Rather, HELCOM and the Helsinki convention were both *outcomes*, reflecting the perceived potential for mutual gains among the countries along the Baltic Sea rim.

As briefly touched upon above, the Baltic Sea Joint Comprehensive Environmental Programme (JCP) was initiated in the aftermath of the Ronneby Conference held in Sweden in 1990. This program was endorsed by the Ministers of Environment (i.e. the head of the relevant Ministry) in all countries adjacent to the Baltic Sea at a conference held in Helsinki in April 1992 (HELCOM 1993b:i). The need to also involve international financial organizations at an early stage was clearly appreciated among the Baltic Sea countries. In the drafting of the program, the World Bank (IBRD), the European Bank for Reconstruction and Development (EBRD), the European Investment Bank (EIB), and the Nordic Environment Finance Corporation (NEFCO) influenced the process of creating a workable scheme. The HELCOM Programme Implementation Task Force (HELCOM PITF) was created with the assignment to coordinate the implementation of the program. However, the implementation of environmental abatement projects was from the beginning seen as a responsibility for each individual country.

The JCP aimed to restore the ecological balance of the Baltic Sea basin in a long-term perspective. This would be achieved through (1) an identification of necessary measures to be taken, and (2) an elaboration of an action plan. The estimated time needed for the implementation of both phases was 20 years (that is, the goal should be achieved in 2010). The action program consisted of four main areas: (a) policy and institutional reforms, (b) institutional strengthening and human resource development activities, (c) infrastructure investments, and (d) coastal lagoons and wetlands management (HELCOM 1993b: Chapter 1). In the process of creating a potential for implementation, a list of 124 so-called *Hot Spots* was made by the HELCOM PITF.[188] The major part, about three quarters, of these 124 prioritized areas were located in the countries in transition. The most urgent cases were labeled *Priority Hot Spots*. The total number of Priority Hot Spots was 49, all situated in the former socialist countries in the south-eastern part of the Baltic Sea region. The total cost for the program for the full period (1993-2012) was estimated to be approximately 150 billion SEK (Sida 1996c: 2). It was pointed out by HELCOM as well as by the contracting parties that a prerequisite for a successful implementation of the JCP was that the countries in transition were provided with external financial resources in one way or another. In 1993, for example, a high level conference was held in Gdansk, Poland, which was exclusively targeted at resource mobilization (HELCOM 1993). Almost all of the 150 billion needed was to be allocated to the action program item Investment activities. According to the estimation of HELCOM PITF, 97 percent of the resources needed would be used for investments. The largest

[188] Originally, 132 Hot Spots were identified, but in 1994 eight of them were removed. The removed Hot Spots were all located in the north-western part of the region, four in Finland and four in Sweden (HELCOM 1994). In the early 2000s, nine additional Hot Spots were removed. As the criteria for the selection were expected to change, however, new Hot Spots would possibly be identified (SNV 2000b).

category of environmental investments would be in the area of municipal waste water treatment and combined municipal and industrial waste water treatment (72 billion SEK). Due mainly to the comparably small size of the Baltic States, however, the resources needed in these countries were much smaller than in Poland and Russia. The estimated cost for the upgrading of waste water treatment ranged from about 800 million SEK in Estonia to 1,500 million SEK in Lithuania.

In order to increase the incentives for the implementation of actual investments, HELCOM institutionalized a procedure of assigning different contracting parties specific areas of responsibility (i.e. making it possible for each contracting part to become a *lead party* in a specific environmental issue-area under the JCP charter). Sweden adopted the lead party role for the issue-area of combined municipal and industrial waste water treatment (HELCOM 1994:2). Sweden had a vast experience of waste water treatment due to the extensive construction of plants in the 1960s and 1970s. The waste water treatment industrial sector, however, experienced overcapacity when the need for additional plants diminished. In other words, Sweden could offer a considerable capacity and expertise in the area of waste water treatment investments, and it was in this issue-area Sweden concentrated most of its resources allocated to the JCP. Up to the end of 1995, a total of 400 million SEK had been distributed through the JCP process, and of these 350 million SEK were spent on waste water treatment (Sida 1996b).

The Swedish investment assistance within the waste water issue-area was in most cases undertaken in collaboration with other countries or international, financial institutions. Most lenders and other providers of financial resources required that a certain percentage of the investment was financed locally, e.g. IBRD required that the recipient country should cover at least 20 percent of the cost. The Swedish share varied between 10 and 30 percent (Sida 1996b). When the project was implemented, the administrative construction was in most cases in the form of a tri-partite contract between Sida, the ministry of the recipient country, and the local water purification plant. It was also common to engage a Swedish, local plant in a so-called *Twinning agreement*, in order to utilize technical expertise.[189]

At least in principle, the Swedish policy was to honor the so-called Polluter Pays Principle (PPP).[190] In the case of waste water treatment, this meant that the consumers should pay the full tariffs for clean water. In this way, Sweden would minimize its own costs for investments abroad. The rationale behind the Swedish aid has been to make it possible to upgrade a regular investment to one more adapted to Western environmental requirements, thereby reducing harmful emissions to the recipients. It has been acknowledged that due to the considerable economic difficulties in the former socialist countries, it was not reasonable to demand that the plants should meet strict environmental standards and at the same time be financed purely by do-

[189] The twinning agreement between Kaunas Water Company, Lithuania and Stockholm Water Company is described in Danielsson (1998). A more general description of Swedish environmental twinning agreements in the Baltic Sea region can be found in Falk (1996).

[190] For analysis over the adherence to this principle in the HELCOM process, see Mickwitz (1998)

mestic resources (Sida 1996c). Local funding should in principle finance the waste water treatment plant, but the "extra" resources needed to make it meet high environmental standards could be provided exogenously.

2.1. Swedish waste water treatment support within the JCP

Estonia

The most extensive project in the area of waste water treatment in Estonia differed from those in Latvia and Lithuania, where the focus was on large city facilities. The water purification and waste water treatment project in the Haapsalu/Matsalu area was part of a more comprehensive environmental initiative, where the aim was to protect the considerable natural, cultural, and historical assets of the area. The city of Haapsalu is located on the Haapsalu peninsula in the Baltic Sea. Despite the city's relatively small size (approximately 15,000 inhabitants), considerable environmental pressure was exerted on the sensitive ecological systems in the region. About two thirds of the population were connected to the local waste water treatment plant, but since only mechanical processing was undertaken, considerable amounts of nutrients were discharged into local wetlands and to the Baltic Sea in the early 1990s.

A joint project supported by Sweden was begun in 1996 and included an upgrading of the existing waste water treatment plant and specific measures to protect the wetlands. The project was designed as a model for other smaller Estonian towns with similar environmental problems (Sida 1996b: 10). Total investments cost were estimated at approximately 47 million SEK, and Sweden covered about one quarter of this. Other co-financiers were the Estonian government, the IBRD, and Finland. The Water Works Company of Haninge, Sweden, entered a twinning agreement with their local counterpart in Haapsalu.

Latvia

The Swedish support to the waste water sector in Latvia was directed to the three relatively large cities of Riga, Daugavpils, and Liepaja. In each of these cities, Sweden started large assistance projects targeted at the waste water treatment sector. Riga, the capital of Latvia, has approximately 900.000 inhabitants. The municipality of Riga was the largest, single contributor of nutrients to the Baltic Sea among the Baltic cities (Sida 1996b: 13). In the mid-1990s only about half of its inhabitants were connected to the sewage network of the municipality (SNV 1996:23). The total investment need in the waste water sector of Riga was calculated to be approximately 600 million SEK, with the Swedish contribution being about 40 million SEK. The largest part of the financing was expected to come from the EBRD, EIB, and from local funding.

The city of Daugavpils is situated on the river of Daugava, from which it gets all its fresh-water. Since the river of Daugava passes through Russian and Belo Russian territory before it reaches the Bay of Riga, the control over upstream pollution has been very limited. Although about 90 percent of the population was connected to the sewage network, purification was limited due to the inadequacy of the purification plant. Additionally, the poor quality of the sewage network made leakage legio, which threatened the quality of the groundwater supply. The Swedish assistance was focused on an upgrading and an increase in capacity of the purification plant. The project was initiated in 1996. The Swedish financial contribution was approximately 22 million SEK, while the complete project cost was about 160 million SEK. The Latvian government, the municipality of Daugavpils and the local water works company, IBRD, EU/PHARE, NEFCO, and Finland covered the remaining part of the investment costs.

The third large project supported by Sweden concerned waste water treatment in Liepaja, the third largest town of Latvia. Like Riga, Liepaja is situated near the Baltic Sea coastline. Coastal waters outside the city were severely polluted. The situation in the Liepaja area was further aggravated by the remnants of the large Soviet Union military base. The full extent of the environmental degradation of the former military base area was unclear in the mid-1990s. Most of Liepaja's inhabitants (approximately 85 percent) but only about 50 percent of the industries were connected to the sewage network at the beginning of the 1990s. The waste water treatment plant, however, only received the polluted water from half of the inhabitants. The remaining part went through the network, but was discharged directly through the lake of Liepaja and into the Baltic Sea. Large investments were needed in the sewage network, as well as in the malfunctioning waste water treatment plant. The project commenced in 1995. Total costs were estimated to be approximately 150 million SEK of which about one third was provided by Sweden. The other sources of financing consisted of the Latvian government, the municipality of Liepaja, the IBRD, NEFCO, and Finland. A twinning agreement was reached between the Water Company in Norrköping, Sweden and the municipality in Liepaja.

Lithuania

Sweden supported projects to upgrade the sewers and waste water treatment plants in three large Lithuanian cities: Kaunas, Klaipėda, and Šiauliai. Prior to the initiation of the project in Kaunas, more than 30 percent of the population was not connected to the sewage network, and for those connected, only mechanical cleaning was performed. The total cost for the upgrade was estimated to be approximately 730 million SEK. The Swedish part of the financing was approximately 40 million SEK, while local funding, EBRD, EU/PHARE, NEFCO, and Finland covered the remaining part. The water purification plant of Stockholm (Stockholm Vatten) was engaged in a twinning agreement for the project that started in 1991 (Lariola & Danielsson 1998).

Klaipėda is a city of about 250,000 inhabitants and is located on the Baltic Sea coast. The coastal waters of the area have been severely polluted due to the lack of effective cleaning in Kaunas, as well as from other pollution sources along the Nemunas and Neris rivers. Although most of the inhabitants were connected to the sewage system, the purification plant was gravely overloaded. Therefore, the construction of a new plant was started in 1995. Of the total cost of approximately 160 million SEK, the Swedish share was about one quarter. The Lithuanian government, local funding, the IBRD and Finland, made up the shortfalls. The Malmö Water Works was engaged in a twinning agreement, together with the local

Works was engaged in a twinning agreement, together with the local company Klaipėda State Water Supply Enterprise.

Šiauliai is located close to the Baltic Sea coastline. All waste water from both industries and the municipality in the early 1990s was discharged into the Kulpe River, which flows into the Lielupe River. The outflow of the Lielupe River into the, at that time, badly polluted Bay of Riga. The fact that the River Lielupe crosses the border into Latvia before it flows into the Baltic Sea, makes the environmental control complicated. In order to enhance coordination and avoid potential conflicts, the Lielupe River Basin Commission was created in 1993. A comprehensive mapping of the sources polluting the river was achieved with Swedish assistance. Since only about 70 percent of the inhabitants in Šiauliai were connected to the municipal sewage network and the purification plant was in rather bad shape, a comprehensive upgrade was seen as important and in 1996 a 154 million SEK project was started. A twinning agreement was reached between the local water works and the municipalities of Kristianstad, Sweden and Trondheim, Norway. The Swedish financial contribution was estimated to be approximately 32 million SEK, while the Lithuanian government, the municipality of Šiauliai, the IBRD, Finland, and Norway covered the residual costs.

2.2. Common features of the supported JCP projects

The most important common characteristics of the Swedish environmental support within the JCP framework were the selection of the waste water sector as the primary target of assistance, the concentration on locations close to the Baltic Sea, and the practice of twinning agreements. The available expertise and the over-capacity of Swedish companies were important factors behind the Swedish decision to take on the Lead Party role in this sector. The large investments needed in this area did not come as a surprise to Sweden. The poor quality of the plants and networks in the Baltic States was well known. By adopting the Lead Party role it became possible for Sweden to gain considerable influence over one of the most important factors affecting the environmental quality of the Baltic Sea. In addition to this environmental opportunity, Swedish investment companies and consultants were given a chance to expand their markets. The concentration on investment locations close to the Baltic Sea was obvious, since these projects have been undertaken within the JCP framework. However, the relatively large size of the Swedish contributions should be noticed.

All the projects supported within the JCP were designed to include a Swedish counterpart, in most cases a public company, through the practice of Twinning agreements. These agreements served different functions. First, Twinning agreements were designed to make lower level collaboration easier than what is usually the case in interstate interactions. Specific technical problems and procedures can be addressed directly and in cooperation with the counterpart in the donor country. Second, and somewhat more indirectly, different kinds of technical solutions and procedures found suitable for different reasons in Sweden influenced the design of the facilities in the recipient country. Swedish companies could in this way achieve a favored market situation in the future.

Although Sweden was the Lead Party for investments in waste water facilities, the relatively low grade of activities concerning the other parts of the JCP should be noted. Few major projects in, for example, the program on non-point source pollution headed by Poland had been implemented. The most important reason behind this was probably the difficulties inherent in, for example, the reduction of agricultural run-off (HELCOM 1993a). What were needed in this issue-area were in most cases not large investments in new facilities, but rather changes practice. In other words, no easy solutions were available to these kinds of problems, neither in the more affluent countries in the West, nor in the countries in transition. Another important reason, however, was the considerably lower pressure from business interests concerning this issue-area, since large investments in most cases would not be part of the assistance project. The design would rather place its focus on for example comparably small pilot projects' facilities, where different solutions adapted to the particular geographical location could be tested.

3. BITS FIELD OF ACTIVITY BEFORE 1995

The National Board for Investment and Technical Support (BITS) was a considerably larger provider of resources to CEE than the main Swedish foreign aid authority, SIDA, during the period before the reconstruction of SIDA in 1995. Since the establishment of the BITS programs in 1989, the appropriation of resources increased steadily from approximately 60 million SEK to 233 million in the budget-year 1994/1995. The annual contribution to projects aiming at environmental improvement varied between approximately 25 and 80 million SEK in the period 1990-1994 (Stockholm Environment Institute 1994:9). It should be noted, though, that Poland was the only recipient of support in the first year. Subsequently, this support was decreased and increasing amounts were allocated to the Baltic States. In the period 1990-1994, a total of approximately 20 million SEK was directed to each of the Baltic States for environmental and energy projects.[191] At this stage, the individual projects were still rather small, which was reflected in the comparably large number of separate undertakings: 26 (Estonia), 16 (Latvia), and 15 (Lithuania).[192]

Like SIDA, BITS did not take an active part in the actual project implementation. When a project proposal from the recipient country was accepted, BITS tried to find a suitable implementing company, organization, or university. The partner BITS was looking for was preferably of Swedish origin. In this way, Swedish national interests could be adhered to at the same time as recipient needs were addressed.

The most important support from BITS in the first couple of years was in the areas of waste water treatment and energy conservation. The assistance resources were at his time generally not spent on investments, however, but on pre-feasibility stud-

[191] Not including 14 million SEK, which were spent on joint, Baltic environmental and energy projects.
[192] Not including 15 joint, Baltic projects.

ies and other types of preparations. A considerable part was used for the procurement of services from Swedish consulting agencies. The difference between SIDA and BITS should be recapitulated. SIDA had been working almost exclusively with very poor countries, while BITS had its primary experience from cooperation with more "advanced" recipient countries. In fact, in the early years of BITS in the first half of the eighties, BITS was explicitly instructed not to make any commitments in countries where SIDA was active. Later on, this distinction became difficult to uphold, and BITS took part in projects in certain SIDA program countries, e.g. Namibia, Zimbabwe, Botswana (Petersens 1996). This was an important reason why the bulk of the resources made available for the CEE countries were channeled through BITS rather than SIDA. The choice of BITS as the main implementing agency gave the Swedish CEE assistance a distinct flavor of a cooperative approach rather than that of traditional aid giving.

4. SIDA GETS THE MAIN RESPONSIBILITY

In the summer of 1995, BITS was incorporated into the new institutional structure of Sida.[193] The Department for Central and Eastern Europe (Central- och Östeuropaavdelningen; Sida-Öst) was made responsible for all cooperative activities with this region. Many of the selected projects were undertaken within the area of waste water treatment and water purification. These, however, generally were selected within the HELCOM Joint Action Programme, which has been addressed above.

4.1. Investments in the energy sector

Except for fresh- and waste water projects, the most important environmental initiatives by Sida were taken in the energy-sector and on issues related to air quality.[194] The energy-sector in the former CEE countries was characterized by "...over-consumption, low productivity and extensive environmental side-effects." (Sida 1996a) The political reforms since independence had in some ways increased the problems in the energy-sector. The collection of charges was, as a general rule, not efficient and it was often not politically possible to increase tariffs. This forced the producers to increase their lending. The relatively low tariffs did not provide the consumers with sufficient incentives to economize on their energy use. Due to the poor level of payment by the consumers, investment needs were often not addressed, resulting in malfunctioning facilities. Sweden did, as most other Western states in the region, see it as necessary to provide external assistance in order to avoid excessive social costs during the transformation to market-oriented production.

In the area of energy policy and regulation, three different projects were undertaken

---

[193] As mentioned above, SWEDECORP, Sandö U-Centre, and SAREC were also made part of the new Sida.

[194] The financially most extensive undertakings in the energy-sector have been within the Climate Convention framework, under the administration of NUTEK.

at a total cost to Sweden of approximately seven million SEK (Sida 1996a: 18).[195] Sida financed an advisor on energy issues in the Ministry of Economy in Tallinn, Estonia. In a similar way, an energy expert responsible for the coordination of donor activities in the IBRD Office in Riga, Latvia, was financed. The Swedish company Swedpower was contracted to undertake a sector study of electrical power in Latvia.

Sida allocated about nine million SEK to the issue-area of institutional support in the energy sector (Sida 1996; 1997). The local District Heating Companies in Tartu, Estonia, received support for the rehabilitation of their facilities. Support on project coordination, procurement etc linked to IBRD financing procedures was granted to the Ministry of Economy in Estonia. A Swedish company, Svensk Byggledning AB, has provided consultancy services related to building standards etc to a building company in Estonia and to the Institute of Architecture and Construction in Lithuania. Finally, an investigation of peat deposits was undertaken in Lithuania.

A relatively large part of the Swedish assistance administered by Sida was devoted to prefeasibility studies and consultation support in connection with international financial organizations (such as IBRD, EBRD, and the EIB). Sida allocated about 100 million SEK to this kind of assistance. About a quarter of this amount was used for consultancy services and the rest for credit (Sida 1996b). In Pärnu, Estonia, consultancy support to the IBRD Energy Program for District Heating Rehabilitation was granted. The project included prefeasibility studies for boiler conversion. In Tartu, Estonia, and Jelgava, Latvia, similar projects were initiated and received Swedish support. In Lithuania, a prefeasibility study on energy saving improvements in apartments and public buildings was undertaken by the Swedish company SWECO AB. Sida covered the cost for the prefeasibility study, while it was expected that IBRD would finance physical investments.

4.2. Energy demonstration and educational projects

Several of the supported initiatives were designed as demonstration projects. The rationale behind this form of commitment has primarily been on the one hand to provide examples of possible solutions and on the other hand as a method of transferring technological knowledge. A total of approximately 20 million SEK was allocated to the Baltic States by Sida in this way. In the Estonian cities of Tartu and Tallinn, pilot projects were initiated to find the most effective measures to save energy in residential blocks. Also in Tartu, an application and demonstration of Swedish standards in district heating equipment was undertaken. The Swedish Company Fjärrvärmebyrån implemented this project, which was financed by Sida. Finally, regarding demonstration projects in the energy sector, a model project on small-scale energy systems was implemented in Skangali, Latvia. This project was implemented in cooperation between the Swedish Enterprise SWECO AB and the local counterpart Latvenergo, and included procurement and installation of electrical supply and heating systems equipment.

---

[195] The following description of energy projects covers the period until Nov. 1996.

A number of projects in the category of environmental education were undertaken. Approximately 15 million SEK was allocated by Sida to this kind of assistance. On the issue of efficiency in district heating, training activities were undertaken in Estonia and in Riga, Latvia. Swedish companies and their local counterparts jointly administered these courses. At the Kaunas University of Technology, a course for student teachers on methods of energy saving in buildings was financed by Sida, and implemented by the Swedish enterprise F. Larsson Konstruktion AB. Support was also given for the establishment of an information center at Kaunas University (Sida 1997). Finally, Sida financed a natural gas seminar in Lithuania for senior personnel in gas distribution companies. The Swedish power company Sydkraft implemented this project.

On a joint Baltic basis, Sida financed two different projects in the studied period. The Swedish power company Vattenfall held a training course on power system operation, with the aim of reducing inefficient power production in the Baltic States. On the issue of energy saving in the industrial sector, a training course was held by the Swedish part ÅF Energikonsult Syd, where senior officials from the industrial sector and from the Ministries of Energy in the three Baltic countries took part.

4.3. Common traits of Sida-financed energy projects

It is quite clear from this brief description of Sida-financed projects in the energy sector, that the initiatives selected had some characteristics in common. In most cases, they were relatively small projects with relatively modest individual costs. As a rule, the implementing Swedish part was a private corporation, although in some cases it was a public company. The implementing actor was either a construction company or a consultancy bureau. Since it was an explicit rule Sida was obliged to follow, these actors were always of Swedish origin. In many cases, the strategy was to assist the potential recipients in achieving relatively large loans from international financial organizations such as the World Bank or EBRD. The support from Sida was not given in the form of direct investments or physical equipment, but rather as assistance in planning, prefeasibility studies, and education. The only exceptions to this rule were the pilot projects. This distribution of support, however, was to change. In an evaluation conducted by Sida itself in late 1996 on the request of the government, it was suggested that investments "...should be a regular component in the Sida support [program]." (Sida 1996:20).

It should finally be noted, that the strategy chosen was in most cases to achieve maximum effect in terms of forthcoming investments in a relatively short time. This generally meant that enough support was given in order to achieve a high probability of getting the project started. The instrument to accomplish this was either to finance the mandatory pre-feasibility studies, or to facilitate the Baltic States' contacts with other potential donors or financiers.

5. ASSISTANCE ADMINISTERED BY OTHER AGENCIES

Several other agencies were involved in the Swedish environmental assistance to the

Baltic States in addition to the main agencies BITS and Sida. The Swedish Environmental Protection Agency (Naturvårdsverket; SNV) had a long experience of cooperation with the CEE countries. Funding of projects from this agency rose from approximately six million SEK in 1992/93 to almost 20 million in 1994/95.[196] The main areas of assistance were capacity building and institutional support, with special attention given to water protection and waste management, nature conservation, and monitoring of transboundary air pollution (Stockholm Environment Institute 1994:16).

Swedish collaboration with CEE furthermore encompassed several universities and individual research projects. The Swedish University of Agricultural Science (Sveriges Lantbruksuniversitet; SLU), however, had a special position. In 1993, SLU was granted 25 million SEK for exclusive use within the Baltic Agricultural Run-off Action Program (BAAP). The BAAP was a part of the HELCOM JCP, and its objective was to reduce agricultural run-off to the Baltic Sea from the Baltic States, Poland, and the St. Petersburg and Kaliningrad regions. The creation of small model agricultural units in combination with training and advisory programs were the main mechanisms within the BAAP initiative to achieve a reduction of run-off.

One of the main tasks of the Swedish Institute (Svenska Institutet; SI) was to promote transnational, cultural and academic cooperation. It was stressed by the government that countries adjacent to the Baltic Sea should be prioritized (SNV 1996:28). The SI did not fund any larger, environmental projects. Rather, its task was to provide grants to advanced research contacts between Swedish researchers and their co-workers in CEE. In addition to these joint, academic undertakings, SI also supported initiatives taken by central and local government organizations, schools, trade unions and the like. SI was not involved in any direct contacts with the CEE counterparts. The implementing actors exclusively handled these contacts themselves. In 1994, a total of approximately 20 million SEK had been provided by the SI for environmentally related initiatives between Swedish and Central and Eastern European co-workers.

5.1. NUTEK and the Swedish Climate strategy

A few Swedish organizations that emphasized technological/commercial initiatives were actively involved in the cooperation with the CEE countries. The National Board for Industrial and Technical Development (Närings- och teknikutvecklingsverket; NUTEK) was the main authority in Sweden responsible for technological innovations. Ever since the first oil crisis in 1973, a central task for NUTEK had been to stimulate different kinds of energy saving initiatives. In 1993, the agency was given a special assignment in the Programme for an Environmentally Adapted Energy System (EAES). This program was specifically targeted at the CEE countries. Although the primary aim was to make energy pro-

[196] About half of this sum was earmarked for activities within the Baltic Sea Joint Comprehensive Environmental Programme (JCP).

duction more efficient, the conversion of production processes in most cases had important and beneficial effects on the environment. As has been briefly touched upon above, Sweden provided assistance to the CEE countries within the Framework Convention on Climate Change (FCCC), which was signed by most countries in Rio De Janeiro in 1992, and came into force in March, 1994. All Nordic countries signed and ratified the Convention. The Climate Convention established some opportunities for so-called Joint Implementation investments. According to Article four:

> The developed parties and other Parties included in Annex 1 ... may implement such policies and measures jointly with other parties and may assist other Parties in contributing to the achievement of the objective of the Convention..." (Cited in Nordic Council of Ministers 1996:29)

Since all industrialized countries, including the Baltic States, who had signed the Climate convention were listed in Annex 1, this was applicable in the case of Sweden and the Baltic countries.[197]

NUTEK was given almost 280 million SEK within the EAES for climate-related initiatives. These resources were to be used on a lending basis, for the financing of projects in the Baltic States, the St. Petersburg and Kaliningrad regions, and in Poland. At the end of 1996, a total of approximately 50 projects had been initiated, of which 31 had been completed (Sida 1996a: 21). Of the completed projects, 12 were undertaken in Estonia, 14 in Latvia and 5 in Lithuania. The over-all rationale behind the Swedish CEE energy program was defined according to a general cost-benefit approach. A given amount of financial resources invested in the countries in transition, rather than in Sweden would achieve a greater reduction of emissions (Government bill 1992/93:179). The magnitude of the difference in cost-efficiency between investments in Sweden and the Baltic country was, however, difficult to estimate. Different approaches resulted in rather diverse outcomes, depending on the factors and counterfactuals used in the calculus. According to an expert report by the Nordic Council of Ministers:

> Investments to reduce emission levels in the Baltic countries would yield a *somewhat* higher cost/benefit ratio than corresponding investments in the Nordic countries. The difference is, however, *not significant* (author's italics, Nordic Council of Ministers 1996:67).

The environmental gains from these kinds of investments were hardly disputed, although their magnitude was difficult to estimate adequately. From the perspective of the Baltic States, it was clear that their emission levels would be greatly reduced,

---

197 The developing countries were excluded from Annex 1, since implementation in these countries of reduction of emission levels would probably require some kind of alternative financing schemes, possibly with grant elements. The construction of such financing schemes was at this time not yet agreed upon. It deserves to be mentioned that the decisions reached at the conference in Rio 1992 included an understanding that the emissions of carbon dioxide should be reduced in all countries taking part in the process. The legitimacy of counting contributions through investments on a grant basis aimed at CO_2-reductions *in other countries* was still highly controversial at the Climate Conference held in Nov. 1998 in Buenos Aires, Argentina.

while the reduction from a Nordic perspective would be considerably more modest. The entailing decrease in sulfur emission would, however, have beneficial effects, especially in the local area surrounding the emission source (Nordic Council of Ministers 1996). The Swedish government stated that the resources allocated for this purpose should be used for investments intended to reduce emissions from point sources, such as power plants, district heating, distribution networks etc (Government bill 1992/93:179). The primary aim would, in other words, not be to achieve a reduction of emissions from diffuse sources.

About half of the projects consisted of boiler conversions and the other half of upgrades related to distribution of energy and increased energy efficiency in buildings. In most cases, the environmental gain came from both increased efficiency and the use of less harmful inputs, such as conversion of boilers fueled with oil shale to bio-fueled heating plants in Estonia. The conversion of boilers seemed to have had a considerably higher efficiency in terms of cost per reduced unit of carbon dioxide. It was estimated that the reduction of one unit of CO_2 required about three times more resources, compared to boiler conversions, if the investments were made in the distribution networks and approximately eight times more if they were made in energy savings in buildings (Sida 1996a).

The main part of the resources allocated to NUTEK within the EAES program was used for commercial lending to the owners of the facilities. A smaller part, however, was used on a grant basis to cover costs in Sweden related to the implementation of the projects.

5.2. Common characteristics of the NUTEK-sponsored projects

The activities administered by NUTEK differed from the approach adopted by Sida in some important aspects . In similarity with BITS, NUTEK's experience came from the implementation of comparably advanced technical systems, and not from foreign assistance projects. The undertakings within the Swedish climate strategy were in many ways influenced by this experience. The support was given in the form of investments in technologically advanced facilities with relatively low levels of ambition regarding policy discussions, institutional support, and development. Despite the fact that the task given to NUTEK within the Climate Convention framework was separate from the developmental activities undertaken by Sida, the representatives in the Baltic States often had difficulties in keeping the activities of the two agencies apart:

> ...very little has been done to explain the climate intent [of the NUTEK assistance], to define and account for additional costs, criteria for the crediting of emission reductions vis-à-vis the involved parties, and how high-level agreements on project implementation shall be formulated. For this reason, the program has been perceived as a part of the Swedish development assistance (Sida 1996a: 22).

5.3. Nuclear safety

The Swedish Nuclear Power Inspectorate (SKI) was in early 1992 given earmarked financial resources for cooperative initiatives with the CEE countries. The most important target was the Ignalina nuclear power plant in Lithuania. The RMBK reactors in Ignalina were of the same type as the ones in Chernobyl, albeit of a later generation, and were regarded as substantially less safe than those in the West. A joint program for the improvement of short-term safety in the short-term was formulated. It was run under the auspices of VATESI, the regulatory nuclear authority of Lithuania, and financed by Sweden (200 million SEK), the Nuclear Safety Account (350 million SEK), and to some extent by domestic funding.

The nuclear safety support was from the beginning a delicate balancing act between improving the safety in the existing plant on the one hand, and – by doing this – postponing the eventual dismantling of an inferior quality plant. The Swedish initiatives to support investments in measures to improve safety were criticized by other donors (Connolly & List 1996). The Swedish response to this critique was to emphasize the need for short-term security enhancement, since not only Lithuania, but also neighboring countries, was dependent on the plant for energy supplies. An abrupt closing of Ignalina was therefore not a realistic alternative. In the spring of 1999, both reactors at the Ignalina Nuclear Power Plant were still in use and Lithuania was using nuclear power for 80 percent of its electricity needs. In August 1999, however, the Lithuanian government declared that one of the reactors would be closed before 2005. This closing of one reactor would, according to the government, cost approximately 21 billion SEK, and a complete closedown of the power plant would cost approximately 34 billion SEK (TT-Reuters/Norrköpings Tidningar, 10-09-1999:12).

In addition to the projects supported by SKI, the Swedish Radiation Protection Institute (SSI) was commissioned by the government to undertake measures for radiation protection primarily in the Baltic States and the north-western part of Russia. Specific means were given to SSI from the Ministry of Foreign Affairs for this assignment. The aim was on the one hand to improve long-term protection (by e.g. developing safety awareness), and on the other hand to produce an emergency action plan in case of a nuclear accident.

5.4. Trade promotion and financial assistance

The most important agency regarding the promotion of international trade was the Swedish International Enterprise Development Agency (SWEDECORP), prior to the reconstruction of Sida in 1995. As mentioned above, this agency was created in 1991, when the Import Promotion Office for products from Developing countries (IMPOD), Swedefund, and the industrial branch of Sida were amalgamated. The main task of SWEDECORP in relation to the CEE countries was to assist the establishment of Swedish companies in the Baltic countries and to improve the export potential of the CEE countries. SWEDECORP, however, did not conduct any projects specifically targeted at environmental improvement. Rather, the environmental component was to incorporate environmental concern into the projects supported on a general basis. The financial means at the disposal of SWEDECORP was between 40 and 80 million SEK annually. These resources were used for risk capital investments. In order to ensure local participation, SWEDECORP was obliged to find local co-financiers. According to an evaluation made by the Swedish Foreign department in 1994, and similar to the experiences of BITS, this had been troublesome (Utrikesdepartementet 1994b: 110).

In 1993 Swedefund was detached from SWEDECORP, and became a government-owned limited liability company (Stockholm Environment Institute 1994:13). The main task of Swedefund was to support joint ventures between Sweden and the CEE countries. The assistance was in the form of concessional loans, as well as in equity capital. The selected projects included a few environmental undertakings, such as for example in the forestry sector of the Baltic States. In other cases, Swedefund worked under the instruction to include environmental considerations in its operations. The Swedish government allocated approximately 40 million SEK annually from the foreign aid budget to Swedefund in the 1991-1995 period. It should be noted, though, that "...significant amounts of the CEE-support have not been put into use, awaiting proper projects." (Utrikesdepartementet 1994b: 111)

On a joint Nordic initiative, the Nordic Council of Ministers established The Nordic Environment Finance Corporation (NEFCO) in 1990.[198] Due to the significant credit risks in the CEE countries, the Nordic Investment Bank (NIB) had in general not offered its support to projects in those countries. NEFCO, however, was created to meet the need of investment capital in the environmental sector of the CEE countries and was directly linked to NIB. The main concern of NEFCO was environmental investments, but it also financed some pre-feasibility studies in connection with the HELCOM Hot Spots program. Several projects were undertaken with NEFCO as an equity partner, where a private company from some of the Nordic countries and a local agent in the Baltic State implemented the project (NEFCO 1996). The typical investment project was in the field of environmental technology, such as co-financing waste water treatment facilities in Liepaja and Daugavpils, Latvia and in Kaunas, Lithuania (NEFCO 1998). In addition to the provision of equity capital, NEFCO was also mandated to issue loans to environmental investments. NEFCO was in many cases a partner itself. These loans were mainly offered on a commercial basis, but nevertheless were important as a way of providing local actors with loans with reasonable interest rates. The Swedish grant to NEFCO during the 1990-1996 period was approximately 23 million SEK annually (Stockholm Environment Institute 1994:18).

Another initiative at the Nordic level should be mentioned: The Baltic Investment Programme (BIP). Although this program did not have any explicit environmental ingredients, its considerable size meant that its activities nevertheless had significant impact in certain environmental areas. For example, the investments in BIP included the industrial energy saving efforts in Estonia, which gave considerable environmental benefits as a positive side-effect (Nordic Baltic Investment Committee 1995:15). BIP was launched by the Nordic Council of Ministers in 1992 for a period of three years, and was provided with a total investment funding of al-

[198] The Nordic Council also provided smaller grants as technical assistance.

most 600 million SEK. In addition to this amount, approximately 150 million SEK was devoted to technical assistance. The Swedish contribution was approximately one third of the total amount. In a matching policy of the European Bank of Reconstruction and Development (EBRD), an additional 300 million SEK was made available for the investment program. The main purpose behind the program was to "channel capital and technical assistance in order to promote investment in small and medium-sized companies" (Nordic Baltic Investment Committee 1995:2). The countries of Estonia, Latvia, and Lithuania were the exclusive beneficiaries of BIP.

6. CONCLUSION

Before 1995, the administration of the Swedish environmental assistance was dispersed among a relatively large number of agencies. In order to achieve financing of larger projects, Sweden often acted as a co-financier together with other countries as well as with international financial organizations. A large number of Swedish and Baltic NGOs and companies were active in the implementation phase of different projects.

The specific characteristics of the different public agencies tended to affect the selection of projects, as well as their implementation. Prior to 1995, BITS and NUTEK managed the largest part of the assistance. Both agencies had a tradition of cooperation with more "advanced" partners and neither of them possessed any implementation capability of their own. After the reorganization in 1995, most of the CEE assistance came under the responsibility of the Department for Central and Eastern Europe at Sida. This did not, however, change the administration of the assistance in any significant way, since the previous practices typically were transferred from the different agencies to Sida.

HELCOM in general and the JCP in particular served as somewhat of a focal point for the assistance. The JCP did not, however, form the Swedish policy, but was rather an outcome from the perceived need to initiate joint environmental projects. Sweden took the responsibility for waste water treatment as a Lead country, due to its competence, interests, and resources in this sector. The HELCOM institution thereby served to decrease transaction costs, as a relatively large number of projects could be integrated within the JCP.

The Swedish Climate Strategy also became an important part of the Swedish over-all environmental cooperation, since comparably large investments to reduce emissions of greenhouse gases were undertaken in the Baltic States.

In the next chapter, the environmental preferences of the Baltic States will be quantitatively compared with the Swedish selection of aid projects, in order to validate the hypotheses of the model. The more qualitative empirical material presented in the last chapters will furthermore be related to the predictions of the model in a more non-technical way.

CHAPTER 7

ANALYSIS AND CONCLUSIONS

1. INTRODUCTION

The primary question to answer in this study has been whether the Swedish environmental support could have been seen as primarily driven by self-interest. In this final chapter, the model outlined in Chapter two is validated against empirical data on the Swedish project portfolio. The relation between the model and the more qualitative empirical material presented in Chapter three to six will be discussed in more detail. It will be shown, that the Swedish environmental assistance policy towards the Baltic States seems to have had important components of self-interest. The driving forces behind the formulation of the Swedish policy were, however, complex. Different aspects of Swedish self-interest influenced the policy strategy, depending on what type of domestic interests it supported. The model used in this study deals only with what type of project was selected, based on collective good content. Factors such as, for example, whether Swedish rather than Baltic consultants were contracted or if particular projects were supported due to attempts to support Swedish industries have not been addressed.[199] If such factors did in fact influence the formulation of the Swedish aid policy, the hypothesis regarding Swedish self-interest would be further strengthened. It should be noted once again, however, that the ambition has not been to completely rule-out Swedish altruism. It is probable, indeed, that altruistic intentions also influenced the Swedish assistance policy. What will be shown, however, is that although altruism may have been present, it was not the most important factor determining the environmental aid policy adopted vis-à-vis the Baltic States.

Foreign aid based on altruistic grounds would in a general sense consist of projects complying with the preferences of the recipient country. The project portfolio would, in other words, be based according to recipient country gains only. It would also, ideally, be characterized by efficient implementation, since this would be in the interest of the recipient country. In the particular area of environmental support to the CEE countries in the 1991-1996 period, it would focus on the countries or regions with the most substantial needs for external support. Regarding the Baltic States, it would be distributed on a roughly equal basis to the eleven most highly prioritized environmental issue-areas. The support to the Baltic States would not be expected to be correlated with the stake in, for example, the common Baltic Sea, but

[199] The practice of using Swedish, rather than Baltic, consultants has been criticized several times by representatives from the responsible authorities in the Baltic States (Löfstedt 1996).

rather with, for example, the size of the donor country's economy and the needs of the recipient country. In this chapter, the threads from the previous chapters will be drawn together, and it will be shown that the Swedish environmental support to the Baltic States primarily was characterized by Swedish self-interest and a search for mutual gains.

2. CLASSIFICATION OF ENVIRONMENTAL ISSUE-AREAS

In Chapter five, a classification was made of the eleven most highly prioritized environmental issue-areas in the Baltic States. It was assumed that these categories were of equal importance to each of the Baltic States. No area was assumed to be more important than any other. In this section it will be shown that there was a difference between this equal distribution and the areas actually selected by Sweden.

The identified areas were:

- Point source air pollution
- Point source water pollution
- Diffuse air pollution
- Diffuse water pollution
- Solid waste
- Fresh-water resources
- Biodiversity and natural resources
- Landscapes and built-up environments
- Institutional strengthening and education
- Nuclear hazards
- Recreation environment values

In Chapter two, the most important dimension of the characterization of collective goods was, for the purpose of this study, said to be geographical in scope. The geographical scope indicates different environmental effects' propensity to dissipate geographically. The identified prioritized environmental issue-areas will now be ranked according to their collective good content. It will then be shown that the Swedish selection of support areas was biased in favor of those with a high collective good content. Since Sweden only benefits from collective benefits, it has incentives to primarily select such projects. In the perspective used here, geographical scope is assumed to be identical to collective good content. The identified prioritized issue-areas by the Baltic States are assumed to be subjectively equally important. Thus it will be possible to test the first hypothesis, which stated that the collective good content would be larger in the Swedish project portfolio than if the selection of projects had been based solely on the preferences of the Baltic States.

The concept of geographical scope is operationalized by making it dependent on two parameters: *medium conductivity* and *probability of being harmed*. This is a major simplification. The three most important omitted dimensions are related to emission volume, diverging toxicity of different substances, and complex propagating patterns. It might seem reasonable to assume that the volume of emitted substances as well as their toxicity would affect the size of the abatement efforts. It seems, however, that due to lack of precise scientific knowledge, as well as the need for political focal points, the policy outcome has been to identify substances and practices that were expected to be harmful. As the extent of the harm is generally not known with any higher degree of precision, abatement measures cannot easily be allocated according to economic efficiency criteria. Resources have thus been allocated to the identified risk areas without explicitly calculating cost/benefit ratios. As noted above, however, calculations of tentative cost/benefit ratios have been made an integral part of allocation schemes *within* a particular environmental issue-area. Finally, the complex propagation patterns means that it is still to a large extent unknown exactly how an emission at one place ends up as contamination in another. This is not only affected by factors such as wind directions, water currents, and physical geography, but also by chemical and biological reactions depending on the transportation medium. It is increasingly appreciated that most substances do become dispersed, although the propagation pace varies considerably. This lack of precise, scientific knowledge has tended to structure the environmental policy according to ordinal categories rather than to cost/benefit ratios. Taken together, these are the reasons why the omission of volume, toxicity, and propagation patterns are reasonable when operationalizing the model's variables.

The most important geographical scope dimension is related to what medium the environmental effect is propagated through, as different media have different conductivity. Generally, air, for example, has a higher conductivity than water. This means that pollution-related changes are more easily spread and transported further away, ceteris paribus, if they affect air rather than water. In a similar way, water has a higher conductivity than soil. But the potential effect is also influenced by a somewhat less apparent factor, the probability of being harmed. Depending on for example meteorological conditions, emissions of harmful substances to the air might affect Swedish environmental interests very differently. In the case of emissions of limited duration (e.g. radioactive substances from a nuclear accident), the meteorological conditions *at that time* are of crucial importance. It then becomes an issue of establishing probabilities of unfavorable wind conditions. When the emissions are of longer duration (e.g. diffuse emissions of acidic gases), the important factor becomes *average* wind conditions, which is analytically treated as the probability at a particular moment of time of being affected by that particular emission. Over time, Sweden would know it was being affected by such a continuous emission. These two vectors in the geographical scope variable, medium conductivity and probability of being harmed, are used to discriminate between the identified categories of environmental issue-areas.

Most of the categories can be categorized according to the medium conductivity dimension. In some cases, however, it becomes necessary to use the probabilistic vector to discriminate between different categories. When comparing for example Diffuse and Point source air pollution, both effects are transmitted through air. One of the most important sub-areas in the diffuse air pollution category is the emission of greenhouse gases (e.g. CO_2 and NO_x). This kind of environmental disturbance to the atmosphere is less important in the point source category, where emissions of for example acidic substances are more central. These acidic substances, however, do not primarily affect the atmosphere, but they do affect surrounding lakes, sea waters, forests etc. We thus find that the probabilistic vector is larger in the case of greenhouse gas emissions, since such gases affect the general atmosphere irrespective of wind conditions, while average wind conditions determine to what extent Sweden is affected by emissions of acidic substances in the Baltic States. For example, meteorological conditions on average are such that it is more common that emissions are carried eastwards than towards the northwest, thus making the probabilistic vector important when discriminating between the two categories diffuse and point source air pollution.

It is quite clear that it is very difficult to measure medium conductivity and the probability of being harmed with any higher degree of precision. As they are only used as instruments to rank the environmental categories, however, they are in most cases sufficiently precise to allow us to discriminate between the eleven categories. As shown below, though, it has not been possible to positively discriminate between all categories. In these cases the categories have been amalgamated.

The use of the medium conductivity and probability of being harmed is quite straightforward when it comes to physical dissemination of environmental disturbances. These instruments will, however, also be used to sort those categories where the medium of exchange is not purely physical. It may, for example, be argued that benefits from biodiversity are more easily transmitted than those of fresh-water resources, since it is becoming increasingly apparent that biological systems are seldom completely separable. Values emanating from biodiversity might also benefit countries other than the country where the project is undertaken, while increased fresh-water resources in a Baltic State hardly benefit Sweden. Obviously, this method is very pragmatic, but will nevertheless fulfill the needs of this study.

The categories will first be put into three subgroups: Large, Medium, and Small geographical scope. All the categories in the Large group are characterized by a larger geographical scope than those of the Medium group, and the categories in the Small group have the lowest degree of geographical scope. It is clear that there are in fact three different subcategories. There is one subgroup of categories where the environmental impact is on a medium that is more or less shared between Sweden and the Baltic country. The categories of diffuse and point source air pollution, diffuse and point source water pollution, and nuclear hazards affect the Baltic Sea or the common air mass directly. Environmental projects targeted at these issue-areas affect Swedish interests directly, since they have an effect on either the Baltic Sea or the common air mass. Consequently, they thus belong to the Large geographical scope group.

The categories comprising solid waste, biodiversity and natural resources, and institutional strengthening belong to the Medium group. Projects undertaken in these areas primarily benefit the country were they are implemented, but there are also significant effects on other countries.

Finally, in the Small geographical scope group the categories of fresh-water resources, landscapes and built-up environments, and recreation environment values are found. Compared to the categories in the other groups, the effect on Swedish environmental interests is here quite insignificant.

Table 5. The eleven categories are divided into three groups, depending on their geographical scope.

Geographical scope		
Large	*Medium*	*Small*
Diffuse air pollution	Solid waste	Fresh-water resources
Point source air pollution	Biodiversity and natural resources	Landscapes and built-up environments
Diffuse water pollution	Institutional strengthening	Recreation environment values
Point source water pollution		
Nuclear hazards		

The categories in each group will now be analyzed in order to reach a complete ranking order.

2.1. The Large geographical scope group

Diffuse and point source water pollution

In a general sense, the use of the Baltic Sea as a sink for different kinds of environmentally harmful substances means that the potential for further dumping diminishes.[200] Low concentrations of such substances may not be very harmful, but as the concentration increases, the risk of serious environmental damage grows, in some cases not linearly. Where this threshold concentration is located differs depending on the kind of substance and the natural characteristics of the sink. Very low concentrations of for example PCBs and heavy metals may cause environmental harm, while considerably higher concentrations of other substances are needed to cause any detectable damage. The conductivity of the Baltic Sea medium must be regarded as comparably very high. It may seem that the geographical location of water pollution, that is the direction vector, is of pivotal importance when determining the collective good content. Although this is true in some cases, especially in the short term, the pollutants nevertheless tend to be transported to the open sea. Because most outlets are located near the large rivers and along coastlines, much of the pollutants end up in the sea. It now seems clear that the riverine flows of pollut-

[200] Weale (1992) has suggested the term "Common-Sink-Resource" for this kind of resource, where potentially harmful substances are deposited in for example the seas. See also Hueting (1980) and Löwgren (1992).

ants into the Baltic Sea are of major importance. According to P. Stålnacke, for example, the nitrogen and phosphorus loads from rivers "... by far exceed the inputs along other pathways, e.g. atmospheric deposition, nitrogen fixation and point source emissions from industries and urban areas along the Baltic Sea coast." (Stålnacke 1996:57) Other substances are deposited, but may later be released into the sea. Once in the Baltic Sea, most pollutants will in time be widely dispersed, although the initial effects may cause severe disturbances locally. This does not mean, however, that these substances become spread out evenly in the Baltic Sea. Concentrations may very well vary, but when they do, it is more due to biological, chemical, and hydrological factors, than to where they were originally emitted. It is often difficult to know where high concentrations will be attained (Danielsson 1998).

It is clear from several Government bills and other official statements, that in the Swedish political discussions on this topic, the image has generally been that these kinds of environmental disturbances are genuinely collective in their character. Emissions to the Baltic Sea directly or to any of its inflows are therefore classified as the category exhibiting the highest degree of collective good content in this group. When comparing diffuse and point source emissions, the latter in most cases tend to be sedimented to a larger extent, due to the higher local concentrations. This means that in the short term, less will be distributed at a distance. Local effects will in most cases be more pronounced compared to diffuse emissions. Point source emissions thus exhibit a smaller collective good content than diffuse water emissions. Diffuse water emissions are consequently categorized as having larger geographical scope than point source emissions.

Diffuse and point source air pollution

Air pollution propagates through an even more efficient medium than water-borne pollution. Emissions to the air are often dispersed over very wide geographical areas. The direction component, however, has a wider range than emissions to the Baltic Sea. Pollutants emitted to the air in a Baltic country are dispersed very differently depending on meteorological conditions. On average, larger proportions tend to be transported by the wind eastwards than across the Baltic Sea. Depending on the type of substance, the geographical scope varies considerably. For example, reductions of carbon dioxide has its main effect on global warming while for example sulfur dioxide reductions have considerable local or regional effects (Nordiska Ministerrådet 1996). Most of the Swedish assistance in this area has been targeted at measures within the Swedish climate strategy. This means that the primary aim has been to decrease emissions of carbon dioxide. An important secondary effect has been a reduction of sulfur dioxide emissions in particular.

The high degree of medium conductivity indicates that the air category (diffuse and point source) has a very large geographical scope. The geographical scope, however, is not only determined by medium conductivity but also by the probability of being harmed. The probability of being harmed by air pollution is considerably lower than by water pollution, simply because only a fraction of the emissions to the air will affect Swedish interests. This probability varies with type of substance, as mentioned above. On average, the effect on Swedish environmental interests is less direct than emissions to the Baltic Sea, and the air pollution category is therefore ranked second in geographical scope. Diffuse air emissions have a larger collective good content than point source emissions because of the same reasons as diffuse and point source water emissions, referred to above. This means that diffuse air pollution is ranked third and point source air pollution fourth, in terms of geographical scope.

Nuclear hazards

The nuclear hazards category possesses distinctive private as well as collective features. The primary target of the Swedish assistance in this area was to install safety-increasing measures at the Ignalina power plant in Lithuania. The risk perceived as most important was that of a major accident, resulting in large emissions of radioactive substances. Although these emissions could be to a water recipient as well as to the air, the latter was of largest significance due to the high conductivity of the air medium. The probability of harm, however, was like other types of air pollution dependent on the wind directions at the time of the discharge. The main reason why this category has been placed as exhibiting the smallest geographical scope in the Large group is the fact that the toxicity of radioactive substances is highly dependant on concentration and that Sweden would only be severely affected in the event of a larger accident. As the concentration would be highest in the near proximity, the Baltic country would suffer considerably more from an accident than Sweden (there is in other words a considerable private part in the event of an accident). In addition to this, the negative consequences from temporary radioactive emissions had to be reduced by both the probability factor of a major accident and that the wind conditions were favorable from a Swedish point of view. The probability of being harmed is thus smaller compared to substances emitted on a continuous basis.

In Table 6 below, the ranking of the categories in the Large geographical scope group is shown.

Table 6. Ranking of the categories in the Large geographical scope group (the lower the number, the larger the geographical scope, and thereby the collective good content).

	Large geographical scope group
1.	Diffuse water pollution
2.	Point source water pollution
3.	Diffuse air pollution
4.	Point source air pollution
5.	Nuclear hazards

2.2. The Medium geographical scope group

Institutional strengthening and education

The institutional strengthening and education category is quite different from the hitherto described groups. The projects in this area were targeted at increasing the administrative capability of the bureaucracy at different levels, as well as to educate professionals and ordinary citizens on environmental matters. This means that the effect in reference to geographical scope varied depending on the specific aim of the particular project. The beneficial effects on Swedish interests from projects in this category were in many cases indirect, emanating from undertakings primarily targeted at other goals. More efficient tax collection schemes would, for example, probably make the financing of environmental undertakings easier, although the primary aim with the administrative reform might not have been of an environmental nature.

To the extent the effects from administrative reform affect categories of large geographical scope, they will also benefit Swedish interests. These benefits would always be less than those emanating from the large geographical scope groups, since the latter are fractions of the former, although they can still be of considerable importance. The institutional strengthening and education category has on these grounds been placed as having the largest level of geographical scope in the Medium scope group.

Biodiversity and natural resources

The biodiversity and natural resources category is primarily characterized by private gains. That is, most of the gains from such projects would benefit the Baltic State, or some of its physical or judicial persons. It can be assumed that the largest benefits from the national natural resources were obtained by the Baltic country, rather than by Sweden, even if external contractors were allowed to harvest the resources. This is probably true, even if outside actors have taken advantage of the relatively weak administrative system during the time of transformation.

The potential gain from preservation of biodiversity is somewhat different. Due to its potentially important long-term effects and to the inherent difficulties in estimating values in this issue-area, it can be assumed that Sweden, through a strategy of caution, attached some value to preserved biodiversity. Certain effects might have been perceived as "common" and therefore worthwhile to invest in. In addition to this, most ecological systems are not fully separable. Initiatives taken in some of the Baltic States may very well have indirect effects on ecological systems that are also important in Sweden (e.g. migrating birds and fish species). In the vocabulary used here, the geographical scope was quite small, but undoubtedly above zero, and this category has therefore been placed second in the medium scope group.

Solid waste

Solid waste has been categorized as having an even smaller level of geographical scope than the above-mentioned groups. Regarding non-toxic wastes (e.g. nutrients) the important effect was on the nearby watersheds. Toxic solid wastes such as for example heavy metals and chlorinated organic substances could have important effects in distant areas, but also in this case, the immediate environmental threat concerned local effects. Local fresh-water resources as well as biota may suffer severe hazards. Ultimately, these substances may reach the sea, but then in low concentrations, since most of them generally have been deposited in the ground or in river and lake sediments. The solid waste category has thus been placed as exhibiting the smallest level of geographical scope in the medium scope group.

In Table 7, the ranking of the categories in the Large geographical scope group is shown.

Table 7. Ranking of the categories in the Medium geographical scope group (the lower the number, the larger the geographical scope, and thereby the collective good content).

	Medium geographical scope group
6.	Institutional strengthening and education
7.	Biodiversity and natural resources
8.	Solid waste

2.3. The Small geographical scope group

The categories of landscapes and built-up environments, recreation environment, and fresh-water resources are all characterized by a very limited geographical scope. Landscapes and built-up environments are resources owned by the nation and its inhabitants and may be protected and maintained or wasted. But it is also true that extraordinary landscapes and buildings in some sense represent a value to other countries too. They may for example represent unique views or construction knowledge that can be shared, and consequently harbor some, albeit small, collective good content.

Recreation environment is also characterized by a very small geographical scope. These kinds of environmental resources are generally not treasured for their uniqueness value in themselves, but rather as assets to be used by the people of the particular country. Foreigners may also use some of these facilities or natural areas, but then it is on terms stipulated by national authorities or private citizens of the country. In other words, the collective good content is very low.

The final category, fresh-water resources, has in similarity with the other two categories in this sub-group a very small geographical scope. It should be noted, that this does not mean that fresh-water resources cannot be adversely affected by other factors, carried by media of larger conductivity (e.g. solid waste). But these effects are captured in other categories. Here, the issue is about the use of the actual resource of fresh-water. As the use of this resource can be restricted quite easily, the transmission across borders is almost zero. To put it plainly, plentiful resources of fresh-water do not benefit Sweden, since the water will be used either by nationals or exported abroad.

Since all the three categories in the Small geographical scope group exhibit very small collective good contents, no attempts have been made to rank them. This is partly due to the small differences in geographical scope, and partly due to the low number of joint initiatives taken in these categories. As it is not desired to stretch the use of the analytical instrument too far, the three categories in this Small scope group will henceforth be treated as only one category, which is labeled Landscapes, recreation and fresh-water. It should be noted, though, that all categories are kept intact when the data are analyzed statistically, since all eleven groups have been defined exogenously. It has not been possible to determine how to rank these three groups, but they do exist since they have been independently based upon expressed preferences in the Baltic States (Chapter five). They are each potential issue-areas at the same level as the other eight categories. In order to ensure validity, the ranking order will be analyzed statistically where the last three categories are placed in such a way that the correlation is the weakest.

Putting together all the categories, the following ordinal list is achieved.

Table 8. Ranking of the categories in the Large, Medium, and Small geographical scope groups (the lower the number, the larger the geographical scope). The last three categories Landscapes and built-up environments, Recreation environment, and Fresh-water resources have been amalgamated into one category.

Ranking of the nationally prioritized categories according to geographical scope	
Large scope	1. Diffuse water pollution
	2. Point source water pollution
	3. Diffuse air pollution
	4. Point source air pollution
	5. Nuclear hazards
Medium scope	6. Institutional strengthening and education
	7. Biodiversity and natural resources
	8. Solid waste
Small scope	9. Landscapes, recreation, and fresh-water

3. THE SWEDISH PROJECT PORTFOLIO

In this section, the Swedish project portfolio will be compared with the eleven categories prioritized by the Baltic countries. It will be analyzed whether the collective good content of the Swedish portfolio is higher than that of the priorities expressed by the Baltic States. It should be recapitulated that it is assumed from the perspective of the Baltic State, that all eleven categories have been perceived as approximately equally important.

Classifying the Swedish assistance projects into the eleven different categories described above is in some cases somewhat arbitrary. Some projects have multiple targets, making the construction of categories not perfectly exclusive. The Swedish share of the over-all financing scheme sometimes targeted specific parts of a project, making the Swedish goals diverge somewhat from those of other contributors. The method used here is that the most important project target, as expressed by the Swedish authorities, has been used. There were very few projects without a single main objective, although it may have had several intended or unintended side effects.

The description of the project portfolio selected by Sweden covers the five years between 1991 and 1996. Projects accepted within this period are included even if they were not concluded until later. It should be noted that some of the projects concerned pre-feasibility or similar kinds of studies. In some cases, they were followed by implementation support at later stages. Such implementation projects have not been included, when there was no positive decision before the end of 1996. Projects with a total Swedish contribution of less than 100,000 current SEK have not been subjected to analysis, nor have projects administered and financed by NGOs. In several cases, NGOs played important roles in project implementation. In most of these cases, however, state authorities provided the largest part of the financial resources. Such cases have been included.

No comprehensive and complete compilations of all Swedish environmental support projects to the Baltic States have been available. Although there is a database administered by the Ministry of Foreign Affairs covering Swedish cooperation with the CEE countries, this was not found to be a complete list of all environmental projects, including their objectives and financing. Therefore, a broad array of sources has been used in order to achieve a compilation that is as valid as possible. The most important sources have been the above-mentioned database, documentation from the Swedish Environmental Protection Board (SNV), external studies, and personal communication with public administrators in the Baltic countries and in Sweden.

A large collection of evaluations of particular projects was at our disposal. Most of these are of relatively poor quality, however, and have therefore not been used for the compilation. In a meta-analysis of all evaluations in the period 1989 to September 1994, it was for example found that: "The last objective [to improve the conditions of the natural environment] has not been commented on or even mentioned in any of the evaluations or reports…" (Andante Consultants AB 1994:52)

In Table 8, the number of projects and level of Swedish commitments are shown for the eleven categories most highly prioritized by the three Baltic States.

Table 8. Classification of projects according to collective good content category.

Category	*Estonia*		*Latvia*		*Lithuania*	
	N	*Amount**	*N*	*Amount**	*N*	*Amount**
Diffuse water pollution	1	4.5	0	0	5	9.1
Point source water	5	12.1	5	84.8	4	50.0
Diffuse air pollution	1	5.4	3	53.3	1	3.7
Point source air pollution[1)]	11	286	11	142	13	230.3
Nuclear hazards	4	9.4	8	17.0	14	53.8
Institutional strengthening and education[2)]	4	12	6	12.3	9	21.0
Biodiversity and natural resources	0	0	0	0	0	0
Solid waste	1	0.24	0	0	3	1.5
Landscapes, recreation, and fresh-water	1	1	1	18.9	0	0
Total	28	330	34	328	49	366

* All amounts in current Million SEK.

1) A large number of the projects in this category concern boiler conversions and similar energy efficiency measures. These projects have been placed in the point source air pollution category, although they also have other beneficial environmental effects.

2) Some of the projects included in the institutional support subset concern administrative support to specific environmental sectors. Despite this, they have been categorized as general environmental institutional support, since the benefits may also spread to other sectors.

Sources: Berg (1995), Utrikesdepartementet (1996), Ministry of Foreign Affairs of the Republic of Lithuania, Aid Coordination Secretariat (1995), Ministry of the Environment of Estonia (1996), Stockholm Environment Institute (1994), SNV (1996), Gromov (1996), Government budget bills 1990-1996, personal communications with officials in the environmental administrations of the Baltic countries.

3.1. Collective good content in the Swedish portfolio

In Diagram 1 below, the amount allocated and number of projects in each category is shown. On the X-axis, the content of collective good aspects is depicted.

Diagram 1. The different categories of supported projects classified according to amount, number of projects, and place on the ordinal scale of collective good content. (Abbreviations: PS: Point Source, NR: Natural Resources).

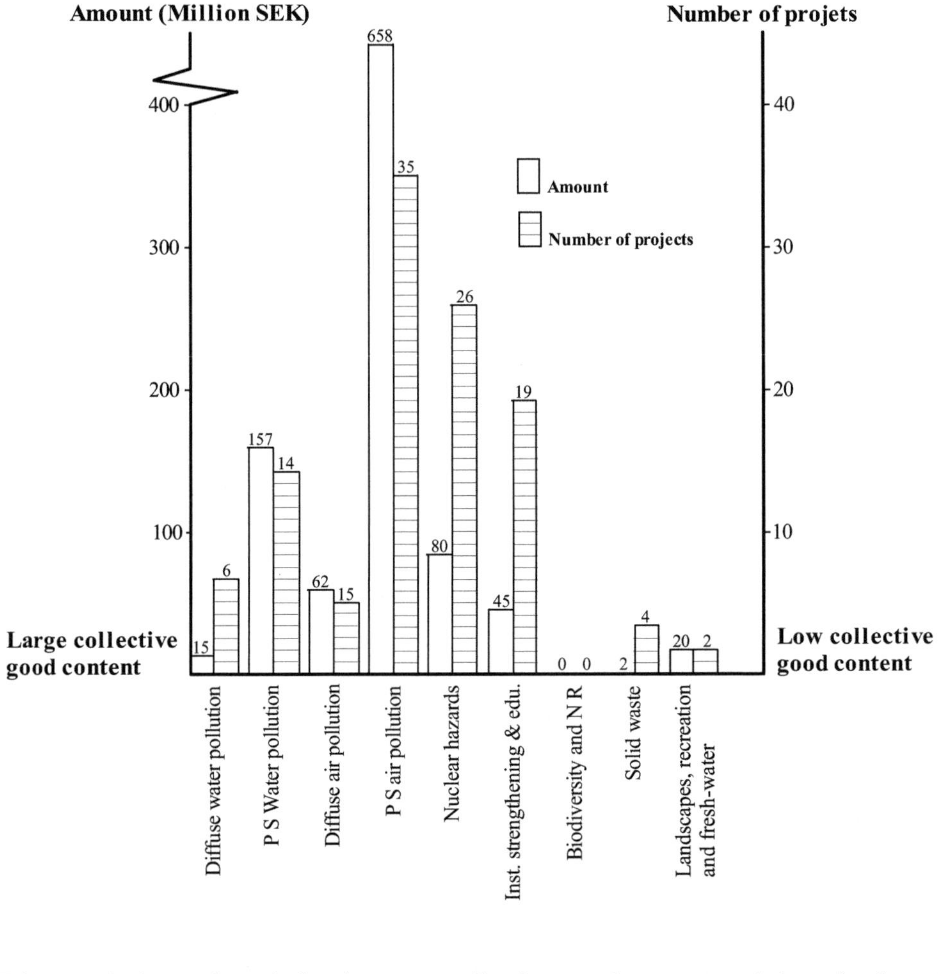

Diagram 1 shows the relation between collective good content and the selection of Swedish support projects. Compared with the assumed equal distribution that had been ideal from the perspective of the Baltic State, however, there is no significant (p=0.05) difference between these sets in relation to either disbursed amounts or number of projects.[201] The rather low p-values, 0.081 and 0.131 respectively, however, make it reasonable to analyze this issue further. It should be recapitulated that all eleven categories have been used when testing for rank correlation. As the ranking of the categories landscapes and built-up environments, recreation environment values, and fresh-water resources is not known, the ranking order resulting in the weakest correlation has been used.

One factor might have made the correlation weaker than it should have been. As can be seen from Diagram 1, Swedish support to point source air pollution abatement was quite extensive. Most of this support was targeted at the reduction of greenhouse gas emissions. The reduction of these emissions was an integral part of the Swedish climate strategy, and has been counted by Sweden as a reduction of *Swedish* emissions. As was discussed in Chapter six, Sweden chose to reduce emissions in the Baltic countries as a complement to reductions in Sweden, since the marginal efficiency in abatement projects was considerably higher in the Baltic States. This means that these investments were more self-interested than what would have been the case if the reductions had not been credited to Sweden. By investing in the Baltic countries Sweden could satisfy both domestic interests and those of the Baltic States. The rationales presented in the official documentation, however, were almost exclusively based on the goal to reduce "Swedish" emission of greenhouse gases. In other words, the self-interest element in this category was probably higher than indicated by the ranking order in Diagram 1.

In the remainder of this chapter, the implications of these findings will be further discussed. The more extensive empirical observations presented in Chapters 3-6 will be drawn upon, to make the picture more complete.

4. STABLE SOCIAL PREFERENCES?

The description of Swedish foreign assistance started with the early policy formulated most concisely in the Government bill of 1962. In this bill, it was emphasized emphatically that Swedish foreign aid should be targeted at the poorest countries and that no domestic, or other less noble interests, should be allowed to influence the direction of Swedish aid flows. In the latter part of the 1960s, however, this began to change. The volume of the aid had grown considerably. Because of its increasing size, it became more exposed to the interests of other actors. The changing political climate and the downturn of the economy also contributed to the first tendencies towards a dichotomization of Swedish aid. Over the following years, part of the aid was targeted at primarily alleviating poverty and analphabetism, while the other was used to reach mutually beneficial cooperation with "more advanced" countries. In parallel with this development, Swedish international environmental cooperation became more pronounced. It became increasingly apparent that different kinds of environmental disturbances did not stop at the borders of a country. Therefore, and

[201] Kendall's tau is 0.420 (p=0.081) for the correlation between ranking and amount spent in a two-tailed test. Corresponding numbers for number of projects are 0.381 (p=0.113).

in order to improve environmentally harmful situations and prevent future dangers, transnational agreements and cooperation were conceived as essential.

At first sight, it may seem that the Swedish preferences in this issue-area have changed considerably. From being defined as almost entirely altruistic to a dual characteristic of altruism as well as of self-interest. Although this would not devalue the validity of the model, since it is only applied on the data from the period 1991-1996 when the preferences seemed surprisingly stable, it is still somewhat disturbing. If the preferences did change, the general applicability of the assumption of stable and socially homogenous preferences would be refuted.[202] It is neither the intention, nor possible, to substantiate the stable and homogenous preferences assumption, but it would still be interesting to see whether it is congruent with the data presented in this study or not.

As described in Chapter four, the opinions of the Swedish political parties on the issue of the CEE support were very homogenous. This was not the case regarding earlier Swedish foreign assistance in general, as described in Chapter three. Rather than drawing the conclusion that Swedish preferences changed, it seems plausible that other individuals and interests than those determining the CEE program influenced the general aid policy. As also described in Chapter three, the traditional foreign aid policy of Sweden was to a large extent influenced by proponents both within and outside the government who strove for a continued focus on alleviating poverty. This means that a collective set of preferences could not easily be found for this period, since this set would be the result of a "game" between the actors with diverging preferences. It is consequently not certain that the preferences did change. It seems more likely that the balance of power between the proponents of altruistic aid and those emphasizing joint gains shifted in favor of the latter. It should be noted, that even though there have always been influential individuals propagating altruistic aid, this does not mean that *their* preferences were altruistic. Their urge for Swedish assistance targeting LDCs might have been based on altruistic preferences, but they might also very well have been based on self-interested preferences, where individual strategies served to increase salary, status, position, or similar. To confirm this suggestion, however, an in-depth study of the key actors of that time would be required, which is out of the scope of this study.

In conclusion, it is clear that the changed Swedish foreign aid policy does not necessarily imply changed individual preferences. Consequently, the assumption of stable and homogenous preferences cannot be refuted. The homogeneity part of the assumption cannot be refuted despite diverging strategies by the sub-national actors, since identical preferences in different structural positions (for example being a member of the government, in opposition or outside the parliament) might very well make different strategies optimal.

---

[202] For a defense of the assumption of stable and similar preferences, see for example Becker (1977).

5. SELF-INTEREST AND JOINT GAINS

The main suggestion presented above was that the Swedish selection of environmental projects seems to have been compatible with the assumption of self-interested preferences. Projects with a high collective good content were selected to a considerably higher degree than projects benefiting only the Baltic country. It is important to emphasize, however, that this *does not* mean that the interests of the Baltic States were sacrificed, when compared to a hypothetical situation without Swedish assistance. The model described in Chapter two clearly showed that both actors had the opportunity to choose not to cooperate, which would mean that no joint projects would have been implemented. Despite the skewed project selection, the net gains from the joint projects exceeded what would have been possible to achieve by the Baltic States themselves. Otherwise, the Baltic States would have abstained from accepting any joint projects. If compared to completely altruistic Swedish assistance, on the other hand, it is quite clear that the utility to the Baltic State would have been greater. First of all, the Swedish assistance would then not have been tied. The procedure of tying aid can only be beneficial to the recipient country under very specific circumstances (i.e. when the value *to the recipient country* of the provided goods or services is larger than their market price). In all other cases, the recipient country would be better off getting the assistance in cash and deciding the optimal resource allocation for itself.

It would, furthermore, have been preferable to the Baltic country if the assistance had been less focused on areas with high collective good content. Although these issue-areas were all part of the prioritized items in all three countries, so were for example management of solid waste, landscapes and built-up environments, and biodiversity. As described in Chapter two, the model predicts not only that Sweden will tend to support projects with a high collective good content, but also that considerable domestic resources will be diverted to these kinds of projects.

From the model in Chapter two, we know that for the domestic project to be preferred to an assisted project, net external support to a particular project has to be larger than the utility difference between the domestic and the assisted project. If this is so, the Baltic country prefers the assisted project. The external support has been at least 25 percent of the total costs in most projects, and in some cases considerably larger, such as when additional donors were engaged in the same project. This makes it clear that there is a significant potential for utility gains by accepting external assistance, especially when external assistance is a large part of the total project cost. For the domestic project to be competitive, the utility difference between the two projects must be significant.

We turn now to the Swedish choice of strategy. For Sweden to prefer an assisted project rather than a domestic one, the collective gain from a project in the Baltic country minus the gross Swedish contribution must be larger than the net gains from a project implemented in Sweden plus the collective good share from the project implemented in the Baltic country. It is assumed that if Sweden chooses not to contribute to the project in the Baltic country, the latter will use the domestic resources in some other project, which might have some collective good content. It is likely, however, that the collective benefit emanating from the alternative domestic project undertaken in the Baltic country if an assisted project is not implemented would be comparably small. Since the joint projects have been shown to be biased towards those with high collective good content, it would be expected that the Baltic State favored issue-areas with low collective good content when the financing was domestic. It might therefore be assumed that when Sweden is considering whether to implement a joint or a domestic project, the collective share attributable to Sweden from the alternative project undertaken in the Baltic State is not substantively affecting the Swedish decision in a substantial way.

It should be noted, that when Sweden covers part of the cost of a project in a Baltic country, it might not be the only external contributor. First of all, the Swedish strategy has often been to facilitate the creation of joint contracts with other donors. This increased the impact from Swedish support considerably. In the common cases where other donors were engaged in the same project, the relevant choice was between financing a domestic project completely or to cover maybe 25 percent of the costs of a project in the Baltic country. With a given amount of Swedish and Baltic State financing, a larger project could be implemented if additional donors contributed, and a larger project in a given issue-area would benefit Sweden more, all other things being equal. When a lower Swedish contribution is allowed for, a lower utility from an assisted project can be accepted, without making Sweden abstain from cooperation.

The willingness to provide assistance is positively related to the collective good content of the project according to the model presented in Chapter two. This assumption is also valid for other bilateral assistance providers, which means that the preferences of for example Sweden, Finland, and Denmark would be expected to be reasonably similar. This would make it attractive to find project proposals with at least one or two additional collaborating partners, even though a larger number of contributors probably would increase transaction costs. As shown by, for example, Berg (1994), this finding has a clear empirical bearing, since a large number of joint projects have been undertaken, where several assistance providers were collaborating in the same project.

A major mitigating factor that might threaten the realization of these joint gains is the incentives for free-riding. As discussed in Chapter two, there is a risk that the individual actor chooses not to implement a project, hoping that someone else will do it instead. The institutionalization of certain procedures under the regime of HELCOM, however, served to decrease the temptation to free-ride. By assigning Lead Party status to different actors, for example, the primary responsibility was placed on a single actor. Without precluding additional assistance from other donors, this mechanism decreased temptations to free-ride. The standard praxis among the donors to cover the costs for additional environmental adaptation of projects otherwise financed by domestic means or support from other donors or IFIs also decreased the risk of free-riding. The donor country could target environmental disturbances particularly important to it, which made it more likely to provide the support. In a similar way it has been common that Sweden, like most other bilateral donors, invested aid resources in order to facilitate the lending process vis-à-vis IFIs.

If a project favorable to Sweden was financed by an IFI, the invested resources in preparation of the application produced a large impact. Although the selection of projects was considerably more restricted when it had to be accepted by an IFI, this loss could be redeemed through the multiplied impact.

The last hypothesis deduced from the model in Chapter two stated that external assistance tends to pull domestic financial resources from issue-areas with low collective good content. The mechanism causing this was the fact that the recipient country was required to part-finance all joint projects. The Swedish contribution to a particular project has generally ranged between 20 and 30 percent. Even if several external financiers covered part of the project cost, the assistance providers almost always required that the recipient country carried a substantial part of the costs. Consequently, in order to obtain the benefits from the joint projects, the Baltic State was induced to divert large amounts from the state's as well as the municipalities' budgets to the projects preferred by the donors. It seems clear that a large part of the restricted domestic resources available in the Baltic States for environmental investments were "locked up" in the often capital-consuming projects favored by external donors. If several donors are involved in the same project, this means that projects that are rather low on the priority list of the Baltic State might still be preferable to a domestic project, since the choice might be between covering e.g. 30 percent of an assisted project or 100 percent of a domestic project. By implementing these assisted projects, however, fewer resources were available for initiatives in areas not prioritized by the foreign donors.

Large investments were needed in for example the solid waste, environmental awareness, and biodiversity sectors, where the level of external assistance was very modest. This was confirmed by several speakers at the EcoCity99 conference on environmental investments in Latvia and Lithuania held in Helsinki, September 9th, 1999. Mr. Olli Ojala, General Director of the Finland-Latvia Environmental Education Project (FiLEE), confirmed this as a general pattern in the Baltic States, while Ms Aivita Zavadska, Deputy Director, Investment Department, Ministry of Environmental Protection and Regional Development in Latvia and Mr. Saulius Smalys, Head of Environmental Strategy Department, Programmes and Projects Management Division, The Ministry of Environment of the Republic of Lithuania emphasized similar needs in Latvia and Lithuania, respectively. In Latvia, about 40 percent of the resources from the Environmental Fund was allocated to water-related issues (1997), while less than 15 percent was allocated to environmental awareness, accident mitigation, rehabilitation measures, nature protection, and biodiversity *in combination*.[203] Furthermore, almost no investments were made in waste management. According to en evaluation conducted by the European Commission for Europe in 1998, "... substantial improvements [regarding waste management] are needed urgently in all areas." (ECE 1999a: 131) The situation was similar in Estonia. It was quite clear that large resources were spent on environmental issue-areas with high collective content, while initiatives in issue-areas exhibiting lower degrees

[203] According to government sources, as reported in ECE 1999b:42.

of collective good content were in many cases insignificant. Considerable amounts were for instance targeted to curbing air emissions and in 1994, approximately 40 percent of expenditure covered by the Estonian National Environmental Fund were targeted at water-related areas (ECE 1996:61).

It could thus be concluded, that although the utility to the Baltic State from undertaking joint projects was higher compared to the situation if only domestic initiatives had been implemented, it would have been even larger if the external resources had been more evenly distributed among the issue-areas prioritized by the Baltic States.

6. ASSISTANCE DYNAMCIS

It was hypothesized in Chapter two that the amount of Swedish assistance would tend to decrease over time, since the difference between marginal net values of projects implemented in a Baltic country and in Sweden would decline. Projects with very high efficiency in terms of net environmental gain would be implemented first. Eventually, the difference in efficiency and possibly paralleled with decreasing differences in GNP would make assistance obsolete. In Poland for example, this process seems to be almost completed. In the late 1980s and early 1990s, Poland was the largest recipient of Swedish environmental assistance (Stockholm Environment Institute 1994). When the most urgent environmental needs had been addressed and the Polish economy had grown sufficiently to cover the costs of prioritized environmental investments, Swedish assistance was reduced. A decade after its inception, Swedish assistance to Poland was almost completely withdrawn. Since it had by and large served its purpose, neither actor had any reason to let it continue.

The model predicts that when net gains from an assisted project are larger for both actors than net benefits from domestic ones, cooperation will take place. Two factors are important in determining the size of the difference between gains from assisted and domestic projects. First, the declining marginal utility from additional assisted projects makes net benefits smaller. Second, if the economic development in the recipient country is more rapid than in the donor country, the opportunity cost of domestic projects decreases. If the recipient country becomes wealthier, the sacrifice in terms of opportunity cost when implementing a domestic project becomes comparably smaller.

Like all models, this is a great simplification of reality. It is for example likely that the external assistance would be terminated before the gains from an assisted project equal that of a domestic project. Factors such as full control over the facility and national pride probably would make the recipient country abstain from further assistance before the above-described threshold is reached. Neither of these factors is included in the model. Another factor that probably would affect the level of assistance and that is not fully integrated in the model has to do with learning. In the first years of Swedish assistance, the amounts provided were lower than in the last years in the studied period, as described in Chapter four. According to the model, the assistance should be at its maximum at the start, and then continuously decrease as the marginal gain difference between an assisted and a Swedish project diminishes. This partial inertia in the early phases is included in transaction costs, but it has not been defined how these costs vary over time. It is clear from the data, however, that it took 2-3 years before the assistance volume reached a steady rate. According to a number of statements by the government, as well as by the political parties in opposition, the intent has always been to eventually withdraw and let the citizens of the Baltic States cover the full costs. In contrast with the situation in Poland, a decrease in the support level had not been observed by 1999. This is not surprising, since withdrawal difficulties are a very common experience to most foreign aid providers. This was further aggravated by the fact that the economic growth in all three countries slowed down towards the end of the 1990s. When the project concerns an issue that needs continuous maintenance of one sort or another, it is essential for future self-reliance that the domestic workforce is capable of providing this.

7. LEARNING FROM EXPERIENCE AND DECREASING INFORMATION GAPS

A conclusion drawn from the model was that the character of the Baltic project proposals would be based on the expected value from the project as well as on the expected probability of approval. More precisely, the strategy would be to maximize the product of these factors. Unless the Baltic State's information about Swedish preferences is imperfect, the probability of approval would always equal unity, provided that the decision-situation was not such that rational behavior resulted in Pareto-inferior outcomes as in, for example, a single stage Prisoners' dilemma. If the Baltic State knew exactly which project would be accepted, it would always propose projects that were approved. In this case, the selection process in the Baltic State would only be based on environmental utility, with the Swedish requirements as the restriction. It is not likely, however, that the Baltic State possessed perfect knowledge about the Swedish preferences. Therefore, the Baltic State tried to estimate the approval probability of different projects, compare this with net environmental utility, and choose those with the highest product. Although no thorough investigation of all applications has been undertaken, a large number of them have been studied and it seems clear that the diversity of the applications has tended to decrease over time. In the first phase, many applications were clearly not compatible with Swedish support policies and were often quite small in terms of total cost. The initiative often came from more or less private citizens, and often was coupled with commercial or particular issue interests. Over time, however, the proposals less frequently tended to have targets too much out of line with Swedish policies. Obviously, a learning process had taken place, where the Baltic States over time became better informed about Swedish preferences at the same time as Sweden became more aware of potential projects with high collective good content.

In a similar way, Sweden has not been able to accurately predict the outcome of a potential project in terms of collective, environmental net benefits. The general tendency has been to favor larger and more complex projects and to adapt the initiatives more to specific and sometimes diverging issue-areas in the different countries.

Furthermore, the donors' assessment of the recipient countries' institutional capacity and the nature of their environmental problems have over time become more accurate.

This learning process within the administration and political bodies in both Sweden and the Baltic States could probably be modeled by attaching a Bayesian update of beliefs model to the game described in Chapter two.[204] By trial and error, the estimation of the other actors' preferences would then become increasingly close to reality during iterated interactions.[205] It should be noted, that the outcome in terms of the number of joint projects undertaken when the actors' information level increases is unclear. It might seem that increased information would make cooperative behavior more beneficial, but this is not certain. If, for example, the actors perceive the situation as being one of a coordination game where both actors benefit from cooperation with certainty, while it actually is a Prisoners' dilemma, the lack of information would induce the actors to cooperate (Hassler 1997). If they had known that it was a Prisoners' dilemma beforehand, the incentive to cooperate would most likely have been smaller. The construction of such a learning model is, however, outside the scope of this study, where it is only pointed out that there seems to have been a process of learning.

Finally, to further illuminate the Swedish strategy, a somewhat different factor influencing the volume of environmental assistance will be presented. This factor is based on a potential link between geographical proximity and the level of ambition of different donors. This aspect is not a part of the model of Chapter two, but it is based on the same assumptions of self-interested behavior.

8. GEOGRAPHICAL PROXIMITY AS A POTENTIAL EXPLANATION

In the first part of this chapter, the concept of geographical scope was operationalized with the parameters medium conductivity and probability of being harmed. Keeping the same concepts, it seems reasonable to assume that the probability of being harmed would be negatively correlated with geographical distance. It will therefore be analyzed whether geographical distance between the donor and the recipient country has affected the selection of prioritized issue-areas of assistance.

Most other important bilateral donors except for Sweden are located at various distances from the Baltic States. Finland, for example, has Estonia just on the other side of the Bay of Finland, while Denmark and Germany are located closest to Lithuania. The case of Finland diverges somewhat from the other bilateral donors, since Estonia for a long time and on several dimensions (i.e. cultural, linguistic, economic) has been "closer" to Finland than the other Baltic countries. A thorough analysis of the selection of Finnish projects would therefore probably also have to take these dimensions into consideration.

---

[204] For a description of this method, see for example Morrow (1994).

[205] Learning in environmental aid negotiations in the context of Western environmental aid to the CEE countries has been analyzed without using formal modeling. See for example Sjöstedt (1996).

According to the geographical proximity hypothesis Sweden would be expected to have allocated similar amounts to all three Baltic States, while Finland would have allocated most to Estonia. Denmark and Germany would be expected to have transferred the largest amounts to Lithuania. In Table 9 below, the amount granted and number of projects initiated by the bilateral donors is given. The amount granted and number of projects initiated by Sweden is given as reference.[206] It should be noted, that Table 9 is based on Berg (1995) exclusively, in order to achieve a more valid comparison between the listed donors and lenders.

Table 9. Bilateral assistance to Estonia, Latvia, and Lithuania (1990 to spring 1994).

	Estonia		*Latvia*		*Lithuania*	
	*Amount**	*N***	*Amount**	*N***	*Amount**	*N***
Finland	80.4	32	16.7	6	16.6	1
Denmark	57.8	18	29.9	16	43.4	17
Germany	12.5	13	8.3	19	7.1	11
IFIs	739.4	39	101.9	46	289.7	38
Sweden	*135*	*25*	*93.4*	*25*	*204.9*	*25*

* All amounts in current SEK. The following exchange rates have been used in the conversion: 1 SEK equals 8.60 Euro, 1.44 FIM, 1.15 DKK, 8.3 USD, and 4.4 DEM.
** N is the number of supported projects.
Source: Berg (1995).

It is evident from Table 9 that Finland allocated the largest share of the assistance to Estonia, which is what was expected according to the geographical proximity hypothesis. The amount allocated to Latvia and Lithuania respectively, however, was almost identical, and the number of projects small, which makes it difficult to draw any clear conclusions. The number of possible values in the dependent variable (recipient country) is in addition too small to make rank correlation analysis meaningful. No clear pattern can be found in relation to the other donors project portfolios in the dimension of geographical proximity. Germany, for example, has the opposite ranking order compared to what was expected, and in similarity with Denmark allocated the largest share to Estonia, which is located at the greatest distance from both countries. The Swedish allocation to the three Baltic States varies considerably. Lithuania received most, while less than half of this amount was allocated to Latvia.

[206] It should be noticed that the data on Swedish assistance differ in Table 9, 10, and 11. This is due to the different sources that have been used, as explained in the text above.

9. VARYING STAKES IN THE ENVIRONMENTAL STATUS OF THE BALTIC SEA

As a final analytical step in this study, the level of ambition of the bilateral donors will be compared to their individual environmental stakes in the Baltic Sea. It was mentioned in Chapter one that the countries adjacent to the Baltic Sea have coastlines of very different lengths. It will now be hypothesized that the length of the coastline correlates positively with the incentives to invest in projects targeted at the environmental status of the Baltic Sea. Although coast length is a very crude measure of the countries' stake in the environmental status of the Baltic Sea, the large differences in coastline lengths make it interesting to see which country it is that provides the largest assistance when this difference has been normalized for. The potential value from a longer coastline stems from commercial and transportation values, as well as from real estate and recreational values. Although the former categories, where for example water-requiring industries, fish stocks and similar are included, are tangible and easily recognized, the value stemming from the latter category is probably many times larger. Investments in private real estates and different kinds of facilities that are more highly valued because of their location by the sea represent large values. These values might decrease significantly if the environmental state of the Baltic Sea deteriorated further. Consequently, the gains from environmental abatement projects increase with the size of the sectors that are in various degrees dependent on the Baltic Sea collective goods. It should be noted, however, that the marginal value from the length of coastlines probably is decreasing, the smaller the opportunity to utilize Baltic Sea goods, the more valuable they become.

In Table 10 below, the length of the coastlines of the bilateral donors is given.[207]

Table 10. Coastline lengths of the bilateral donors (percent).

Bilateral donor	*Percent of total coastline length*
Sweden	60
Finland	17
Germany	5

Source: Personal communication with the Swedish Meteorological and Hydrological Institute (SMHI), 1999-04-21.

In addition to the comparison between the amount granted by these bilateral donors and their Baltic Sea coast lengths, the amounts granted will now be compared also with the gross national products (GNP) of the donor countries. The larger the GNP, the larger absolute amounts would be expected to be allocated as assistance to the Baltic States. Similarly prioritized items in different national budgets would receive

[207] Denmark has been excluded since adequate data have not been available.

different real amounts depending on the size of the donor country's GNP.

In Table 11 below, total assistance from Sweden, Finland, and Germany has been multiplied with each country's proportion of the total Baltic Sea coast length. This operation gives us the amount each country would have contributed, if the total support had been identical to the actual support, and each country had contributed in strict relation to its coast length. This amount is labeled Coast length Adjusted Assistance (CAA). In the last column, total assistance from all three countries is instead related to the proportion of the sum of the GNP's of all three countries. We thus arrive at hypothetical assistance amounts, if each country had contributed according to their GNP size, and total assistance from all three countries was the same as the actual amount. This amount is labeled GNP Adjusted Assistance (GAA).

Table 11. Total assistance by three donors in relation to Baltic Sea coast length (CAA) and GNP (GAA).[208]

Bilateral donor	*Gross assistance⊠*	*Coast length adjusted assistance (CAA)**	*GNP Adjusted Assistance (GAA)***
Sweden	52.2	50.7	6.4
Finland	13.7	14.4	3.2
Germany	3.4	4.2	59.7
Total	*69.3*	*69.3*	*69.3*

⊠ Data over gross assistance have been taken from Berg (1995), which means that the period 1990 to spring 1994 is covered. The exchanges rates used were: 1 SEK equals 8.60 Euro, 1.44 FIM, 1.15 DKK, 8.3 USD, and 4.4 DEM. All amounts are in current million US$.

* Data over coastline lengths have been obtained through personal communication with the Swedish Meteorological and Hydrological Institute (SMHI), 1999-04-21.

** Data over GNP have been taken from International Bank for Reconstruction and Development (1993; 1995). The average of the GNP data from 1991 and 1993 has been used for the calculation of GAA.

In Diagram 2, the data from Table 11 are presented in graphical form.

[208] It should be noted that the omission of the assistance from Denmark as mentioned above does not affect CAA or GAA, since the total amount used when calculating these values are the amounts contributed by Sweden Finland, and Germany. This total amount has been treated as given, and the CAA and GAA hypotheses only speculate about the proportions contributed by each of the three countries.

Diagram 2. Actual assistance, Coast length adjusted assistance (CAA), and GNP adjusted assistance to the Baltic States from Sweden, Finland, and Germany.

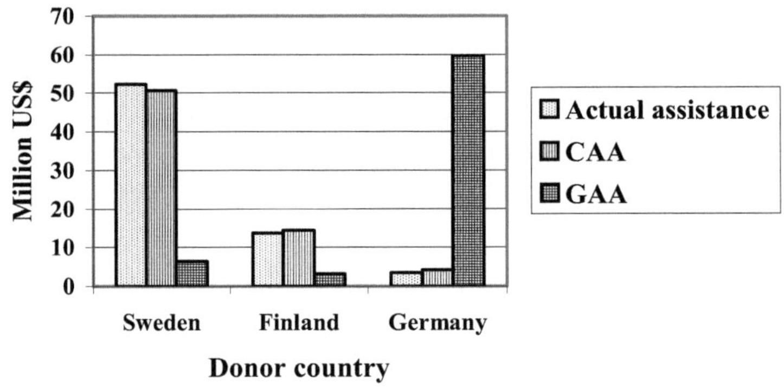

It is quite clear from Table 11 and Diagram 2 that the cost length adjusted assistance corresponded well with actual assistance. The relatively large differences in provided assistance correspond to a similarly large difference in proportion of coastline. It does not seem, however, that the size of the economy had any significant impact on resources spent on environmental assistance to the Baltic States. Sweden and Finland were contributing much more than their share of the total GNP would imply, according to the hypothesized relation between that and assistance described above, and Germany contributed a lot less than what would have been expected in reference to its large economy.

10. SWEDISH ASSISTANCE TO THE BALTIC STATES; RECENT DEVELOPMENTS

10.1. The government policy vis-à-vis the CEE countries

The third Swedish CEE assistance program was about to end in late 1998. A new three-year program had been formulated, and was accepted by the parliament in May 1998. The level of ambition in terms of allocated resources was kept unchanged at a total of approximately one billion SEK per year (Government bill 1997/98:70 p. 4). In the new 2001 program there were, however, some changes worth noticing. The Swedish environmental support was in the 1991-1996 period almost exclusively targeted at the Baltic States and Poland. In the early 1990s, the support to Poland decreased as the Polish economy grew at a considerable pace. Although it had already been stated as desirable, the need to support the north-western part of Russia (especially the St Petersburg and Kaliningrad regions) was further reaffirmed, as was support to Ukraine and Belarus, albeit to a lesser extent.

The over-all aim was to let the development assistance evolve into more traditional cooperative interactions with neighboring countries, according to the SAP government (Government bill 1997/98:70). According to the same document, this aim was about to be realized, and was already " … dominating in the cooperation with some of the [CEE] countries." (Government bill 1997/98:70 p. 7) Most donor countries, however, probably shared this over-all aim. It is seldom intended to keep the recipient countries dependent on foreign aid. In praxis, this has often proved to be difficult (Cassen 1987). Within the CEE program, only the Swedish assistance to Poland has actually decreased. Due to favorable economic development in Poland in the second half of the 1990s, opportunity costs for environmental investments decreased, thus making external assistance less valuable. There is not yet any corresponding development discernible in the Swedish relationship with the Baltic States. Considering that the economic development was rather weak in all three countries in the late 1990s, the opportunity cost of environmental investments will most likely continue to be high also in the first years of the 21st century.

Sweden expressed an intention to change the weight given to different sectors in the new CEE program. Resources should be transferred from general institutional support to, for example, environmental assistance. The size of the environmental component in the support program would thus increase.

In 1999, Estonia was well under way to becoming a member of the EU, as the only one of the Baltic countries accepted among the most highly prioritized applicant countries.[209] Latvia and Lithuania were included in the second round of membership negotiations. Ever since Sweden became a member of the EU, it has given high priority to supporting the Baltic accession to the Union. In the earlier three-year programs, the first phase was primarily targeted at supporting national sovereignty and the second at strengthening democracy and market institutions. In the late 1990s, the third phase – European integration – had become the most important. According to the government, almost 90 percent of the support in the 1998-2000 program was more or less targeted at facilitating the accession to the EU (Government bill 1997/98:70 p. 14). The national so-called pre-accession strategies of the Baltic countries were explicitly supported by Sweden, and Sweden also actively campaigned in the EU for faster membership negotiations with the Baltic States and to increase the level of EU support, primarily through the PHARE program. A closer integration of the Baltic States with Western Europe was seen as valuable to Sweden, not only from an environmental perspective but also from security and trade perspectives.

The content of the Swedish environmental support was in the third program kept along similar lines as in the previous two. The largest part of the investments was as before within the framework of the HELCOM JCP and the regional Agenda 21 initiative (Baltic 21) and mainly targeted at waste water treatment facilities. As the

---

[209] At the end of the decade, the categorization of the EU applicant countries into two tracks with different pace was not fully upheld. Countries in the second group that were particularly successful in their accession strategies were thus allowed to overtake less successful countries in the first group. The date of the respective countries membership approval was, however, in late 2000 difficult to predict.

larger cities in all three countries in most cases now had such projects under implementation, increased attention would in the immediate future be given to smaller, coastal towns and the curbing of diffuse emissions of nutrients from agriculture. Projects aimed at increasing energy efficiency, in production and distribution as well as in consumption, were given high priority, as were different measures targeted at nuclear safety. The energy efficiency projects were still an integral part of the Swedish Climate Strategy. These projects had previously been administered by NUTEK, but were in 1998 transferred to the new Energy Authority (Statens Energimyndighet). Regarding the Ignalina nuclear power plant in Lithuania, Sweden was still of the opinion that it could not be immediately closed down due to "reasons of energy balance" (Government bill 1997/98:70 p. 51). Additional assistance would be given to increase security levels and to strengthen the domestic security authority VATESI.

10.2. Continued political consensus on the CEE policy

Just as during the 1991-1996 period, the political parties in opposition in the late 1990s had only minor objections to the contemporaneous Swedish policy vis-à-vis the CEE countries in general and the Baltic States in particular. The largest of the political parties in opposition, the Conservatives, agreed with the governmental view that assistance would also probably be needed in the foreseeable future (Party bill 1998/99:U802). The focus on nuclear safety, waste water cleaning, and energy efficiency measures was supported by the Conservatives, as was the assistance to the Baltic States in their pre-accession strategies. Like the SAP government, the Conservatives wanted to use the CEE program to support almost exclusively the countries in close proximity such as the Baltic States, the north-western part of Russia, and, to a lesser extent, Ukraine and Belarus. The only critique put forward was regarding the administration of the CEE support. As described earlier, Sida received the main administrative responsibility in 1995. The Conservatives' opinion was that the Ministry for Foreign Affairs ought to have the main responsibility, as it had had before the reconstruction of Sida. According to the Conservatives, the fact that a relatively large part of the CEE assistance was in praxis used under the discretion of the Ministry for Foreign Affairs showed that it would be reasonable to create an administrative unit within that ministry. This unit should be given the main responsibility for the CEE assistance. It would not be appropriate, however, to establish a separate agency for this purpose, according to the Conservatives (Party bill 1998/99:U802 p. 7).

The Liberal Party agreed with the general lines of the governmental CEE policy (Party bill 1998/99:U203). It is interesting to see how distinctly traditional foreign aid policy was separated from the CEE support. The Liberals strongly criticized the tendency to let domestic interests influence traditional foreign aid:

> We oppose the fact that the Swedish self-interest – rather than our opinion on what is actually best for the recipient countries – is allowed to influence the content and formulation of the assistance. Even if Sweden from an international perspective has a more restricted policy such a commercialization of the foreign aid policy must be minimized (Party bill 1998/99:U203 p. 11).

On the other hand, in the same bill it was emphasized that cooperation with the CEE countries was not the same thing as cooperation with countries in the Third World. It is not clear in what way these two types of assistance ought to be different, but the view of the Liberals seems to have been that the peoples of the CEE countries have to be untied from the yoke placed on them by communist rule. Then development would follow almost automatically. By implication, this view was not applicable to the countries in the Third World, where more extensive support schemes were supposed to be needed. The recipient countries of Swedish CEE assistance should be those suggested by the government. The support to Russia ought, however, to be more closely targeted at those regions exhibiting the most promising development in terms of democracy and economic reform (Party bill 1998/99:U203 p. 15).

The general opinion about the CEE policy of the Christian Democrats seems to have been in line with that of the government. No Party bills directly related to this issue were submitted to the parliament in the 1996-1999 period. In Party bill 1998/99:MJ748 the HELCOM JCP was mentioned briefly, and the importance of reducing oil-pollution from vessels was underlined. No changes in the governmental policy on these issues were suggested, however.[210]

The Left Party seemed to agree in general with the government on the formulation and implementation of the CEE assistance (Party bill 1997/98:U41). The perception of the CEE assistance as something separate from the regular Swedish foreign aid policy was apparent. It was clearly the opinion of the Left Party that it expected Sweden to benefit from the support in terms of, for example, increased trade and domestic employment. No references to tied aid inefficiencies were made.

> The Left Party perceive this support as pure investments that will give substantial return both to Sweden and to the recipient countries. Sweden gets back invested resources in the form of increased trade and employment, better environment and security, decreased military spending, and an enhanced life quality for the individual (Party bill 1997/98:U41 p. 2).

The only issue where the Left Party seems to have disagreed somewhat with the government policy was concerning the emphasis placed on assisting the Baltic countries to become members of the EU. Being quite firmly opposed to Swedish membership in the Union, it would perceivably be rather inconsistent to support the Baltic countries wholeheartedly in their approximation process vis-à-vis the EU. The Left Party, on the other hand, did not want to act in the opposite direction. Their suggestion was rather that Swedish authorities ought to assist the CEE countries by compiling an in-depth analysis of the pros and cons of becoming a member of the European Union. Whether to join or not would then be up to the citizens of the CEE countries (Party bill 1997/98:U41).

Somewhat surprisingly, the Green Party did not submit any Party bills referring

---

[210] Within the HELCOM framework, however, the so-called no-special-fee-system was implemented in the summer of 2000. Together with the requirement that every port in the Baltic Sea area should have reception facilities, this system was expected to decrease the incentives for discharging oil at sea (SNV 2000c).

directly to the governmental CEE policy during the 1996-1999 period. Although the environmental status of the Baltic Sea was discussed extensively in for example Party bill 1996/97:JO738, there were no explicit suggestions on changes in the CEE policy.

Taken together, the consensus among the political parties on the formulation and implementation of the CEE assistance program did not change in any significant way during the second half of the 1990s. The relatively small number of Party bills discussing the issue and the few diverging opinions expressed in those bills submitted to the parliament were both signs of a continued high degree of consensus. No critique was raised against the partial change in country selection where less support was given to Poland and more to the northwestern part of Russia (including Kaliningrad) as well as to Ukraine and Belarus. All political parties except for the Left Party supported the increased weight given to assisting the Baltic States in their attempts to become members of the European Union. The latter was not outright against an EU membership for the Baltic countries, but would have preferred a somewhat more neutral Swedish stance on this issue.

The dichotomization of Swedish foreign aid into one part providing traditional aid to LDCs and primarily targeted at alleviating poverty and poverty-related problems, and one part targeted at countries of lower middle-income levels and primarily motivated by perceived mutual gains seems to have been even more institutionalized recently. The support to the Baltic States is in contemporary documentation seldom compared to the traditional assistance policy, but perceived as two almost exclusive categories. Therefore, it does not seem totally unrealistic to assume that the suggestion of the Conservatives to remove the administration of the CEE assistance from Sida and create a new unit within the Ministry for Foreign Affairs will be implemented.

11. SUMMARY AND FUTURE RESEARCH

The analysis of the Swedish environmental assistance to the Baltic States began by describing the early general Swedish foreign aid policy in the 1960s. It was shown that Swedish aid at that time was quite trivial in volume and guided by a clear sense of idealism. The aid was targeted almost exclusively at alleviation of poverty. A decade later, the volume of foreign assistance had become many times larger but had also been increasingly influenced by other than altruistic interests. The sheer size of the foreign assistance, the economic downturn, the oil crises, and the decreasing relevance of the most radical political movements all added to the increasing influence from domestic interests on Swedish foreign aid policy.

At the start of the 1980s, environmental issues were advancing on the political agenda. The complexity of the consequences on the natural environments of for example the Baltic Sea, the air, and animal (including human) life caused by emissions from industry, transportation, and consumption as well as from the modern way of living became increasingly appreciated. This complexity was not only of relevance regarding cause-effect relationships but also concerning geographical dispersion. As many kinds of environmental disturbances have a transboundary character, there were in many cases clear conflicts between the sovereignty of national states and efficient management of environmental issues. An increased level of international harmonization and cooperation was required to safeguard against detrimental outcomes desired by no one.

By the late 1980s, Poland had obtained a significant degree of autonomy from Moscow. Several joint environmental initiatives financed primarily by Sweden were under implementation. Around the turn of the decade the Soviet Union disintegrated and the three Baltic States Estonia, Latvia, and Lithuania regained their independence. This opened up considerably larger opportunities for cooperation among all the countries on the Baltic Sea rim. The Helsinki Commission (HELCOM), created in 1974, served as a focal point for environmental cooperation in the region. The most important environmental hot spots in the region were identified in the 1990 HELCOM Joint Action Programme, as were necessary requirements for the restoration of the Baltic Sea. The priorities of the different member countries in HELCOM were, however, not identical. Therefore, national agendas were compiled, and bilateral interactions initiated. In the spring of 1990, the first Swedish CEE assistance program was accepted by the parliament. In the following three years almost two billion SEK were allocated to assistance primarily targeted at Poland and the Baltic States. Almost 90 percent of this amount was taken from the regular aid budget. About a quarter of the support was used for environmental assistance initiatives.

The most striking feature of the CEE assistance program was its extraordinarily high degree of consent in the parliament. All political parties were outspoken advocates of such a program. The only issue where opinions diverged sharply was on how the program should be financed. Several of the parties in opposition disagreed with the Social Democratic (SAP) government budgetary decision to take almost all of the funding from the regular aid budget. In 1991, the SAP government was replaced by a coalition of non-socialist parties. The new government changed the funding of the program in their first budget, thus restoring the aid volume to the Third world. The SAP did not have any objections to this, since this change had also been discussed within the Social Democratic government. The volume of the support was now increased considerably, from approximately 600.000-700.000 to about one billion SEK annually. In the first three-year program, one billion SEK was allocated to the CEE assistance program. Additional initiatives were, however, taken in several other departments, which means that the total expenditure was about 600.000-700.000 SEK annually. When the Social Democrats returned to power in 1994, they did not change the level of funding of the CEE assistance.

The dichotomization of the Swedish foreign aid where the CEE assistance was perceived as separate from traditional aid considerations was further institutionalized. During the first half of the 1990s, BITS and NUTEK administered most of the assistance. They administered larger investment schemes and technical assistance, while SIDA had the responsibility for NGO cooperation and similar initiatives. Despite the fact that almost the complete administration of Swedish support to the Baltic States in 1995 was placed in a separate unit in the restructured aid agency Sida, the mutual gains criteria were not questioned. The bulk of the assistance was targeted at waste water cleaning and reduction of point source air emissions. Funding to the latter issue-area was primarily given within the Swedish climate strategy, where a reduction of emissions of greenhouse gases was seen to be more efficient if undertaken in the Baltic States rather than in Sweden. These reductions were perceived as "Swedish" in relation to the work within the UN Framework Convention on Climate Change.

In order to evaluate the possible bias in terms of benefiting Swedish environmental interests, the type of projects selected by Sweden was compared to contemporaneous environmental preferences in the Baltic States. It was shown that the most highly prioritized environmental issue-areas in the three Baltic States could be categorized into eleven groups. These groups were not ranked according to their salience, as perceived by Estonia, Latvia, and Lithuania respectively. From the available documentation, the assumption that all these issue-areas were perceived as approximately equally important seemed justified. This is also in line with the experience from most Western states, where important environmental issue-areas are identified, but a precise ranking order of them is seldom specified. Using a non-formalistic game theoretic model, this distribution was then compared with the allocation of Swedish assistance in relation to the eleven issue-areas. The benefit from a typical project in each category was analytically separated into one private and one collective part. The former benefits only the country where the project was undertaken, while both countries gain from the latter. One of the hypotheses generated by the model was that the collective good content should be larger in the set of categories selected by Sweden, than for the optimal set according to the expressed preferences of the Baltic States. Such a difference was found, although the difference was not statistically significant either for amounts provided or number of projects sponsored.[211]

It seems possible that geographical distance was an important factor explaining differences in support volumes, as proximity makes the probability higher that the donor country is within the geographical scope of a particular project. As Sweden is located at approximately the same distance from all the three countries, similar allocations of support to each country would be expected. Although the number of projects was in fact identical in all three countries, the financial amounts varied between 93 (Latvia) and 205 million SEK (Lithuania). Approximately 135 million SEK was allocated to Estonia during the 1991-1995 period. Of the other more prominent bilateral assistance providers, only Finland showed a pronounced tendency to allocate resources according to geographical distance, since Estonia was given more than twice as much as Latvia and Lithuania combined. The latter two countries were, however, assisted with similar amounts, despite the fact that Latvia is closer to Finland than Lithuania. No sign of favoring the closest countries regarding the assistance portfolios of Denmark and Germany was found.

Finally, it has been observed, that there seems to have been a tendency among the key bilateral aid providers to allocate assistance resources in relation to their

---

[211] The correlation for amount was, however, below the $p=0.10$ level, while the corresponding value for number of projects was just over this level.

individual share of the Baltic Sea coastline. The proportion of the total environmental support given to the Baltic States corresponds very well with the respective proportion of coastlines. Sweden provided 52 percent of the total assistance, and has 51 percent of the coastline. Corresponding numbers for Finland were 14 and 14 percent, and for Germany 3 and 4 percent. No tendency to provide assistance in relation to the size of the donor country's economy was found.

The Swedish environmental assistance to the Baltic States in the 1991-1996 period was found to exhibit a tendency to favor joint gains rather than to allocate support resources only in accordance with the priorities of the recipient countries. As has been stated with emphasis, it was not an expectation that the collective good aspects of different environmental issue-areas would be the sole allocation determinant of Swedish aid. As few statistically significant correlations have been found, this expectation was confirmed. Further research in order to establish other factors influencing the particular case of Swedish environmental support to the Baltic States in the 1991-1996 period would be valuable. One such factor would probably be the ability of domestic organizations and corporations to influence the selection of particular assistance projects. Business as well as labor interests certainly have incentives to try to increase the resource flow to specific sectors, industries, and corporations. A further potentially important determinant probably is the governmental tendency to let foreign assistance be influenced by domestic regional development policy considerations.

12. TENTATIVE POLICY IMPLICATIONS

The fact that the Swedish environmental assistance to the Baltic States from the outset has been characterized by a dual emphasis on both joint gains and traditional, altruistic assistance, has resulted in difficulties in measuring its efficiency and legitimacy. When evaluating efficiency in foreign aid programs, it has been common to explore whether positive effects on the recipient country can be detected.[212] If positive effects can be confirmed, the aid project is usually considered successful. This is a very modest level of ambition. Projects with no positive effects whatsoever are probably rare. Regarding international cooperation targeted towards the realization of mutual gains, the efficiency yardstick is different. In these kinds of projects, there are no moral inducements to continue a program that either part finds unsatisfactory. The existence of any kind of positive effects is not sufficient reason to continue a project. The yardstick is here closer to the theoretical concept of opportunity cost (i.e. the gain from a cooperative project has to be larger than the alternative use of these resources could realize). In other words, the efficiency requirements on joint cooperation are considerably higher than on foreign assistance. The dual official rationale behind the Swedish CEE assistance is therefore unfortunate. It might undermine the legitimacy of the assistance program, since it can be argued that it has

---

[212] See for example the report series SASDA (Secretariat for Analysis of Swedish Development Assistance).

considerable efficiency deficiencies if evaluated from a purely international cooperation perspective. If the assumption on the other hand is that the program ought to be guided by traditional Swedish development aid criteria, its legitimacy might be compromised because Swedish interests are allowed to be too influential on the formulation of the program. The reason behind the ambiguity caused by the foreign aid dichotomization is political. By evoking altruistic as well as self-interested justifications, important groups with divergent opinions on this matter can more easily be satisfied. The costs in terms of efficiency losses that will occur from both perspectives are not easily observable. Therefore, the tendency not to specify what underlying principles ought to govern the CEE assistance seems to continue.

Regarding the potential of generalization of this study, it should be emphasized that it is not possible to draw any conclusions regarding driving forces behind other assistance providers' policies, since this study has been exclusively focused on Swedish behavior. The approach and model used, however, is most likely possible to apply also on data from other donor countries. The first issue would then be to establish whether a unitary actor approach is valid. If not, it is necessary to formulate the relevant game at the domestic level. After having identified the prioritized environmental issue-areas in the recipient countries, the application of the model would then be quite straightforward.

REFERENCES

SWEDISH PARLIAMENT BILLS, COMMITTEE REPORTS, AND GOVERNMENT DECISIONS[213]

Government bills

1962: 100
1981/82: 100
1989/90: 100
1990/91: 100; 5
1991/92: 100; 4
1991/92: 150
1992/93: 99
1992/93: 100; 4
1992/93: 179
1993/94: 100; 4
1994/95: 100; 4
1996/97: 1
1997/98: 70

Government decisions

UD96/1834/EC, II: 11
UD96/1835/EC, II: 10
U11, 25
UD2000/276/AF

Party bills

1990/91: Jo 752
1990/91: U203
1990/91: U244
1990/91: U501
1990/91: U516
1990/91: U531
1990/91: U622
1991/92: Jo668
1991/92: U505
1991/92: U506
1991/92: U507
1991/92: U522
1992/93: Jo14
1992/93: Jo631
1992/93: U219
1992/93: U251
1992/93: U255
1992/93: U258
1992/93: U409
1993/94: U304
1993/94: U808
1994/95: U26
1994/95: U30
1994/95: U31
1994/95: U32
1994/95: U33
1994/95: U808
1996/97: U813
1996/97: U814
1998/99: MJ748
1998/99: U41
1998/99: U203
1998/99: U802

Committee reports

1990/91: NU36
1990/91: UU02
1991/92: UU17
1992/93: UU16
1993/94: UU16

[213] All parliament bills, reports and government decisions are originally in Swedish. When these sources have been cited, the author has made the translation.

PRINTED SOURCES

Andante Consultants AB (1994). Development cooperation with Eastern and Central Europe - a review of evaluation results. Stockholm, SASDA.

Abelson, R. P. (1995). *Statistics as principled argument.* Hove, UK, Lawrence Erlbaum Associates, Publishers.

Allan, Pierre & Schmidt, Christian, Ed. (1994). Game theory and international relations; preferences, information and empirical evidence. Aldershot, Edward Elgar.

AMBIO (1990). No. 7.

Andersson, C. (1986). Breaking through; Politics and public opinion in the period of rapid expansion. *Swedish development aid in perspective.* P. Frühling. Värnamo, Fälths Tryckeri AB.

Anell, L. (1986). Images of distant countries: Reflections on the theoretical foundations for Sweden's development cooperation policies. *Swedish development aid in perspective: Policies, problems & results since 1952.* P. Frühling. Värnamo, Fälths Tryckeri AB.

Arrow, K. (1951). *Social Choice and Individual Action.* New Haven, Cowles Foundation.

Axelrod, R. (1984). *The Evolution of Cooperation.* New York, Basic Books.

Bacchus, H. (1990). "Against conditionalization." *Synthese* (85): 475-506.

Balčiūnaitė, Ona (1993). Results of an attempt to determine natural diversity in different landscapes of Lithuania. In Energy, environment and natural resources management in the Baltic Sea region – 4th international conference on system analysis. Report 1993: 653 p. 425-30. Copenhagen, the Nordic Council of Ministers.

Banerjee, Abhilji & Weibull, Jörgen W (1992). *Evolution and rationality: Some recent game-theoretic results.* Stockholm, The Industrial Institute for Economic and Social Research.

Barrett, S. (1994). "The Biodiversity game." Environmental and Resource Economics 4: 111-122.

Barrett, S. (1995). *The strategy of joint implementation in the framework convention on climate change.* UNCTED/GID/10.

Bendor, Jonathan & Swistak, Piotr (1997). "The evolutionary stability of cooperation." *American Political Science Review* 91(2): 290-307.

Berejikian, J. (1995). *Prospect theory and the "Gains Dilemma" in international relations.* Annual meeting of the American Political Science Association, Chicago.

Berg, C. (1995). The environmental support to the Baltic States, Poland and western Russia. Stockholm, Swedish Environmental Protection Board.

Bergström, G. W. (1994). Environmental policy and cooperation in the Baltic region. Uppsala, Uppsala Universitet.

Bernauer, T. (1995). "The effect of international environmental institutions: How we might learn more." *International Organization* 49: 2 (Spring).

\- 1996. Protecting the Rhine river against chloride pollution. *Institutions for environmental aid.* Keohane, R. O. & Levy, M. A. Cambridge, Massachusetts, The MIT Press: 201-32.

Birk, Küllike, Lokk, Elin, Kruglova, Jevgenia & Kaljulaid, Lindu (1993). Microbial pollution in drinking water. In Energy, environment and natural resources management in the Baltic Sea region – 4th international conference on system analysis. Report 1993: 653 p. 273-76. Copenhagen, the Nordic Council of Ministers.

Brams, S. J. (1985). Super power games, applying game theory to superpower conflict. New Haven, Yale University Press.

Bueno de Mesquita, Bruce & Lalman, David (1992). *War and Reason: Domestic and international imperatives.* New Haven, Yale University Press.

Bueno de Mesquita, Bruce (1999). Principles of international politics: People's power, preferences, and perceptions. Washington D. C., CQ Press.

Carraro, C., Ed. (1997). *International environmental negotiations, strategic policy issues.* Cheltenham UK, Edward Elgar Publishing.

Cassen, R. (1987). *Does aid work?* Oxford, Clarendon Press.

Cesar, H. S. J. (1994). *Control and game models of the Greenhouse effect.* Lecture Notes in Economics and Mathematical Systems, 146, Springer Verlag.

Coleman James S. & Farraro, Thomas, J. (1992). *Rational choice theory: Advocacy and critique.* London, Sage Publications.

Coase, R. (1960). "The problem of social cost." *Journal of Law & Economics* III.

Connolly, B & List, M. (1996). Nuclear safety in Eastern Europe and former Soviet Union. *Institutions for environmental aid: Pitfalls and promise.* K. Levy. Cambridge, Mass., The MIT Press: 233-280.

Corell, E. (1999). The negotiable desert: Expert knowledge in the negotiations of the Convention to Combat Desertification (Dissertation no. 286). Linköping, Linköping University.

Corrales, Javier & Feinberg, Richard, E. (1999). "Regimes of cooperation in the Western hemisphere: Power, interests and intellectual traditions." *International Studies Quarterly* (45): 1-36.

Cox, G. W. (1999). "The empirical content of rational choice theory: A reply to Green and Shapiro." *Journal of Theoretical Politics* 11(2): 147-69.

Danielsson, Åsa (1998). *Spatial modelling in sediments* (dissertation no. 178). Linköping, Linköping University.

Denrell, Jerker (2000). "Radical organization theory: An incomplete contract approach to power and organizational design." *Rationality and Society* 12(1): 39-68.

Department of Energy and International Affairs (1993). *The Baltics – Eastern European Series Country Report.* Washington D.C.

Downs, A. (1957). *An economic theory of democracy.* New York, Harper and Row.

Downs, George W & Rocke, David M. (1995). *Optimal imperfection? Domestic uncertainty and institutions in international relations.* Princeton, New Jersey, Princeton University Press.

Dragunas, A. (1995). Country Report: Lithuania. In *Status of the national environmental action programs in Central and Eastern Europe.* The Regional Environmental Center. Budapest, REC: 63-70.

ECE, Economic Commission for Europe (1996). *Environmental performance Review: Estonia.* Number 1. New York and Geneva, United Nations.

- 1999a. *Environmental performance Review: Lithuania.* Number 4. New York and Geneva, United Nations.

- 1999b. *Environmental performance Review: Latvia.* Number 5. New York and Geneva, United Nations.

Edgren, G. (1986). Changing Terms. In *Swedish development aid in perspective: Policies, problems & results since 1952.* P. Frühling. Värnamo, Fälths Tryckeri AB: 47-64.

EEIC, Estonian Environment Information Centre (1996). *Estonian National Environmental Strategy.* Tallinn, Department of Environment.

- 1999. *Keskkond 1997, Estonian Environment.* Tallinn, Ministry of the Environment of Estonia.

Eistere, Inese & Junkurs, Andris (1994). Conservation issues of Liepaja area. In Proceedings of the international conference on oil terminals, shipping and off-shore activities in the Eastern Baltic: Present situation, future prospects, economic and technological appraisal and environmental risk, page 97-99. 11-14 May, 1994. Tallinn, Stockholm Environment Institute-Tallinn.

Elgström, Ole. (1992). *Foreign Aid Negotiations*: The Swedish-Tanzanian aid dialogue. Avebury, Aldershot.

- 1997. *Miljömålet för svenskt bistånd: Beslut och genomförande.* Lund, Lunds Universitet, Statsvetenskapliga institutionen.

Environmental Data Centre (1993). *Keskkond 1992, Estonian Environment.* Helsinki, National Board of Waters and the Environment.

Environmental Protection Committee, Republic of Latvia (1992). *National Report of Latvia to UNCED.* Tallinn, Environmental Protection Committee, Republic of Latvia.

Environmental Protection Department, Republic of Lithuania (1992). *National report of Lithuania.* Vilnius, Environmental Protection Department, Republic of Lithuania.

ERM (Environmental Resources Management Ltd (1998). *National programme for adoption of the Acquis.* London, ERM.

Etzioni, A. (1988). *The moral dimension: toward a new economics.* New York, Free Press.

European Commission (1998). Regular report from the Commission of Estonia's progress towards accession (Global assessment of implementation). Brussels, European Commission.

Falk, H. (1996). Twinning programmes with local authorities in Poland, Estonia, Latvia and Lithuania. Stockholm, Sida.

Fearon, J. D. (1998). "Bargaining, enforcement, and international cooperation." *International Organization* 52(2): 269-305.

Finansdepartementet (1993). Inriktning och samordning av Sveriges samarbete med Central- och Östeuropa. Stockholm, Finansdepartementet.

Fiorina, M. P. (1996). Rational choice, empirical contributions, and the scientific enterprise. In *The rational choice controversy.* J. Friedman. New Haven and London, Yale University Press.

Forsse, A. (1986). With the benefit of hindsight: Reflections of a former aid bureaucrat. In *Swedish development aid in perspective: Policies, problems & results since 1952.* P. Frühling. Värnamo, Fälths Tryckeri AB: 249-260.

Frühling, P. (ed.) (1986). Swedish development aid in perspective: Policies, problems & results since 1952. Värnamo, Fälts Tryckeri AB.

Gautschi, Thomas (2000). "History effects in social dilemma situations." *Rationality and Society* 12(2): 131-62.

Geddes, B. (1994). *Politician's dilemma.* Berkley, University of California Press.

Gil & Folmer, H. (1998). Linking environmental and non-environmental problems in an international setting: The interconnected games approach. In *Game theory and the environment.* H. Folmer & N. Hanley (eds), Cheltenham, Edward Elgar.

Gilpin, R. (1981). *War and change in world politics.* Cambridge, Cambridge University Press.

Goldmann, K. (1978). Det internationella systemet: En teori och dess begränsningar. Stockholm, Aldus/Bonniers.

Gooch, G. (1995). *Territories of environmental concern* (dissertation no 88). Linköping, Linköping University.

Gourevitch, P. A. (1996). "Squaring the circle: the domestic sources of international cooperation." *International Organization* 50(2): 349-73.

Gromov, Allan (1996). Environmental Projects in Estonia implemented with foreign assistance. Unpublished.

Hardin, Garreth (1968). "The tragedy of the commons." *Science* 162: 1243-8.

Haas, K. L., Ed. (1993). Institutions for the earth: Sources of effective international environmental protection. Cambridge, The MIT Press.

Haas, P. M. (1990). Saving the Mediterranean: the politics of international environmental cooperation. New York, Columbia University Press.

Hanley, Nick & Folmer, Hank, Ed. (1998). *Game theory and the environment.* Cheltenham, Edward Elgar.

Harman, G & Thomson, J. Jarvis (1996). *Moral relativism and moral objectivity.* Cambridge, Mass., Blackwell.

Hassler, Björn (1997). Thou should not know too much. When lack of information promotes cooperative behavior: The case of oil-fields in Wyoming, Oklahoma and Texas. In *Managing common resources in local and global systems: Applying theory across scales.* Research report no. 9. Linköping, EPOS.

Hauptmann, E. (1996). *Putting choice before democracy: a critique of rational choice theory.* Albany, N.Y., State University of New York Press.

HELCOM (1988). Declaration of protection of the marine environment of the Baltic Sea area (adopted in 15 February).

- 1993a. High level conference on resource mobilization. Gdansk, Poland.

-1993b. The Baltic Sea Joint Comprehensive Environmental Action Programme: opportunities and constraints in programme implementation (Report no. 49). Helsinki Commission, Helsinki.

- 1994. *Annual Report of HELCOM PITF.* Helsinki, Helsinki Commission.

- 1995. *HELCOM handbook.* Helsinki, Helsinki Commission.

- 1997. List of valid HELCOM recommendations. Helsinki, Helsinki Commission.

Helsinki Commission (1994). 20 years of international cooperation for the Baltic marine environment 1974-1994. Helsinki Commission, Helsinki.

Hellström, T. A. H. (1994). Links between development assistance and donor country exports; the case of Sweden. Stockholm, SASDA.

Henttonen, J. (1996). Cooperation for improvement of the environment in North-Eastern Europe: The Finnish concept. In Löfstedt, R. E. & Sjöstedt, G (ed), *Environmental aid programmes to Eastern Europe: Area studies and theoretical applications.* Brookfield USA, Aldershot.

Heppling, S. (1986). The very first years, memories from an insider. In *Swedish development aid in perspective; policies, problems & results since 1952.* P. Frühling. Värnamo, Fälths Tryckeri AB: 13-27.

Hiltunen, Heidi (1996). Finland's environmental aid to Eastern Europe. In Löfstedt, R. E. & Sjöstedt, G (ed), *Environmental aid programmes to Eastern Europe: Area studies and theoretical applications.* Brookfield USA, Aldershot.

Hirshleifer, J & Glazer, A. (1992). *Price theory and applications.* Eaglewood Cliffs, New Jersey, Prentice-Hall Inc.

Hjorth, R. (1992). *Building international institutions for environmental cooperation* (dissertation no. 81). Linköping, Linköping University.

- 1995. Baltic environmental cooperation; a regime in transition. Linköping, Linköping University.

Hoel, M. & Schneider, K. (1997). „Incentives to participate in an international environmental agreement." *Environmental and Resource Economics* 15: 51-70.

Hopmann, P. T. (1996). *The negotiation process and the resolution of international conflicts.* Columbia, South Carolina, University of South Carolina Press.

Hueting, R. (1980). New scarcity and economic growth. More welfare through less production? Amsterdam, North-Holland Publishing Company.

Hägerhäll, B., Berbeka, K., Kaasik, T., Kallaste, T., Pivoriunas, D., Plate, A., Sukyte, J. (1996). Environmental implications of the EU membership of Estonia, Latvia, Lithuania, and Poland. An assessment of the present situation and the issues and problems involved in the approximation process. Stockholm, Stockholm Environment Institute.

IBRD, International Bank for Reconstruction and Development (1993). *World development report: Investing in health.* New York, Oxford Univ. Press.

- 1995. World development report: Workers in an integrating world. New York, Oxford Univ. Press.

Jacoby, R. (1986). Idealism versus Economics: Swedish aid and commercial interest. In *Swedish development aid in perspective: Policies, problems & results since 1952.* P. Frühling. Värnamo, Fälths Tryckeri AB: 85-100.

Kaempher, William H. & Lowenberg, Anton D (1999). "Unilateral versus multilateral international sanctions: A public choice perspective." *International Studies Quarterly* 43: 37-58.

Kahneman, A. T. D. (1986). "Rational choice and the framing of decisions." *Journal of Business* 59(4): 251-78.

Kallaste, Tiit (1993). Implementation of economic instruments in pollution control in Estonia. In Energy, environment and natural resources management in the Baltic Sea region – 4th international conference on system analysis. Report 1993: 653 p. 197-98. Copenhagen, the Nordic Council of Ministers.

Kallaste, Tiit, Maurer, Küllike & Vesi, Tönü (1996). Country report: Estonia. In *Approximation of European Union Environmental legislation.* The Regional Environmental Center for Central and Eastern Europe (REC). Budapest, REC.

Kallaste, Tiit & Tarand, Andres, Ed. (1998). *Country case study on climate change impacts and adaptation assessments in the Republic of Estonia.* Stockholm Environment Institute Tallinn Centre and United Nations Environment Programme. Tallinn, Vaba Maa.

Kanazawa, S. (1998). "A possible solution to the paradox of voting." *The Journal of Politics* 60(4): 974-95.

Keohane, R. O. (1984). After hegemony: cooperation and discord in the world political economy. Princeton N.J., Princeton University Press.

Krasner, S. D. (1983). "Structural Causes and Regime Consequences: Regimes as Intervening Variables". In *International Regimes.* S. D. Krasner. Ithaca, Cornell University Press.

Krehbiel, K. (1991). *Information and legislative organization.* Ann Arbor, University of Michigan Press.

Kremenyuk, V. A., Ed. (1991). *International negotiation: Analysis, Approaches, Issues.* Processes of international negotiations (PIN) project. San Francisco, Jossey-Bass Publishers.

Kreps, D. M., Roberts, J. & Wilson, R. (1982). "Rational cooperation in the finitely repeated prisoners' dilemma." *Journal of Economic Theory* 2(27): 245-52.

Kreps, David M & Wilson, Robert (1982a). "Reputation and imperfect information." *Journal of Economic Theory*(27): 253-79.

(1982b). "Sequential Equilibria." *Econometrica* 50: 863-94.

Kreps, David M. (1990). *A Course in Microeconomic Theory.* Hempel Hempstead Hertfordshire, Harvester Wheatsheaf.

Kristoferson, Lars (1993. Energy in the Baltic Sea area – Elements of a new situation. In Energy, environment and natural resources management in the Baltic Sea region – 4th international conference on system analysis. Report 1993: 653 p. 19-28. Copenhagen, the Nordic Council of Ministers.

Kudrenickis, I., Jurmalietis, R. & Vircavs, M. (1994). Comparison of oil and renewable fuels: Advantages and disadvantages. In Proceedings of the international conference on oil terminals, shipping and off-shore activities in the Eastern Baltic: Present situation, future prospects, economic and technological appraisal and environmental risk, page 96. 11-14 May, 1994. Tallinn, Stockholm Environment Institute-Tallinn.

Külvik, Mart & Tambets, Jaak (1995). *Key points of the National Biodiversity Action Plan.* June 1995. Tartu-Tallinn, World Bank, Global Environmental Facility, WWF Baltic Programme.

Kuresoo, A., Ader, A., Luigujõe, L. (1994). Concentration areas of non-breeding waterfowl vulnerable to oil pollution in Estonia: Progress report. In *Proceedings of the international conference on oil terminals, shipping and off-shore activities in the Eastern Baltic: Present situation, future prospects, economic and technological appraisal and environmental risk*, page 60-87. 11-14 May, 1994. Tallinn, Stockholm Environment Institute-Tallinn.

Kutas, Saulius (1994). The Lithuanian energy policy. In Proceedings of the international conference on oil terminals, shipping and off-shore activities in the Eastern Baltic: Present situation, future prospects, economic and technological appraisal and environmental risk, page 100-108. 11-14 May, 1994. Tallinn, Stockholm Environment Institute-Tallinn.

Lagerspetz, E. (1997). "Social choice in the real world II: Cyclical preferences and strategic voting in the Finnish presidential election." *Scandinavian Political Studies* 20(1): 53-67.

Lainela, Seija (2000). "Baltic accession to the European Union". *Journal of Baltic Studies* XXXI(2): 204-216.

Lariola, Martti & Danielsson, Birgitta (1998). Twinning cooperation between Kaunas Water Company, Lithuania and Stockholm Water Company. Stockholm, Sida.

Laver, M. (1997). Private desires, political action: An invitation to the politics of rational choice. London, Sage Publications.

Leesment, Liidia & Raud, Riina (1993). The necessity of investigating viral pollution of water environment. In Energy, environment and natural resources management in the Baltic Sea region – 4th international conference on system analysis. Report 1993: 653 p. 287-90. Copenhagen, the Nordic Council of Ministers.

Levy, J. S. (1997). "Prospect theory, rational choice, and international relations." *International Studies Quarterly* (41): 87-112.

Lewin, S. B. (1996). "Economics and Psychology: Lessons for our own day from the early twentieth century." *Journal of Economic Literature* (34): 1293-1323.

Liblik, Valdo & Rätsep, Arvo (1993). The total effects on aerotechnogenic influxes from power generating and cement plants on the north-east Estonian atmosphere pollution level. In *Energy, environment and natural resources management in the Baltic Sea region – 4th international conference on system analysis*. Report 1993: 653 p. 91-102. Copenhagen, the Nordic Council of Ministers.

Lieven, A. (1993). The Baltic revolution; Estonia, Latvia, Lithuania and the path to independence. New Haven, Yale University Press.

Linde, Inga (1996). Country report: Latvia. *In* Approximation of European Union environmental legislation. Budapest, REC.

Lindholm, S. (1970). *U-landsbilden - en undersökning av allmänna opinionen* (dissertation). Stockholm, Stockholm University.

Lindström, U. (1970). Och regeringen satt kvar!: ur min politiska dagbok 1960-1967. Stockholm, Bonnier.

Ljunggren, B. (1986). Swedish goals and priorities. In *Swedish development aid in perspective: Policies, problems and results since 1952*. P. Frühling. Värnamo, Fälts Tryckeri AB: 65-84.

Loftsson, Elfar, Choe, Yonhyuk & Ahlqvist, Anders (forthcoming). *Baltic Barometer.*

Lowry, R. C. (1997). "The private production of public goods: Organizational maintenance, managers' objectives, and collective goals." *American Political Science Review* 91(2): 308-23.

Lundestad, G. (1985). Öst, väst, nord, syd; Huvuddrag i internationell politik 1945-1985. Lund, Studentlitteratur.

Lundqvist, L. J. (1996). Environmental politics in the Nordic countries; Policy, coordination and capacity. Sweden. In *Governing the environment, politics, policy and organization in the Nordic countries*. P. M. Christiansen. Århus, AKA-Print A/S. Nord 1966:5: 259-339.

Löfstedt, Ragnar E. & Sjöstedt, Gunnar (1996). *Environmental aid programmes to Eastern Europe.* Aldershot, Avebury.

Löfstedt, Ragnar E. (1996). A review of Sweden's environmental aid to Eastern Europe: Criticism and possible solutions. In Löfstedt, R. E. & Sjöstedt, G (ed), *Environmental aid programmes to Eastern Europe: Area studies and theoretical applications*. Brookfield USA, Aldershot.

Löwgren, M. (1992). *Being outside the EC environment: Environmental cooperation in the Baltic Area.* The entry into new economic communities: Swedish and Canadian perspectives on the European Economic Community and North America free trade accord, Toronto, Swedish-Canadian Academic Foundation.

Macy, M. (1989). "Walking out of social traps: A stochastic learning model for Prisoners' dilemma." *Rationality and Society* (1): 197-219.

McCubbins, G. C. M. (1993). *Legislative Leviathan*. Berkeley, University of California Press.

Mansfield, Edward D. & Bronson, Rachel (1997). "Alliances, preferential trading agreements, and international trade." *American Political Science Review* 91(1): 94-107.

Mickwitz, P. (1998). Implementation of key environmental principles: Experiences from the protection of the Baltic Sea. Copenhagen, Nordic Council of Ministers.

Milner, H. (1991). "International theories of cooperation among nations." *World Politics* 44(466-96).

Ministry of the Environment, Republic of Estonia (1992). *National report of Estonia to UNCED*. Tallinn, Ministry of the Environment, Republic of Estonia.

- 1996. Environmental projects in Estonia implemented with foreign assistance. Tallinn, Ministry of the Environment, Republic of Estonia.

- 1998. *National Environmental Action Plan – Estonia*. Tallinn, Ministry of the Environment, Republic of Estonia.

MEPRD, Ministry of Environmental Protection and Regional Development of the Republic of Latvia, (1995). *National environmental policy plan for Latvia*. Riga.

- 1996a. National environmental action program. Riga.

- 1996b. State of the environment. Riga.

- 1997a. Environmental and sustainable development projects in Latvia. Riga.

- 1997b. Environmental investments in Latvia. Project Department. Riga.

- 1998. Environmental investment programmes in Latvia. Riga, Investment Department.

MEP, Ministry of Environmental Protection of the Republic of Lithuania (1994). *Environment sector Lithuania - Situation paper*. September 1994. Ministry of Environmental Protection of the Republic of Lithuania.

- 1994b. *Lithuania's Environment: Status, processes, trends*. Vilnius, Ministry of Environmental Protection of the Republic of Lithuania.

- 1995. *National action plan for conservation of biological diversity*. Vilnius, Ministry of Environmental Protection of the Republic of Lithuania.

- 1996. *Lithuanian environmental strategy*. Vilnius, Ministry of Environmental Protection of the Republic of Lithuania.

Ministry of Foreign Affairs of the Republic of Lithuania, European Integration Department, Aid Coordination Secretariat (1995). *Project database – Lithuania.* European Integration Department, Ministry of Foreign Affairs of the Republic of Lithuania.

Ministry of Industry and Energetics, Republic of Estonia (1991). *Energy master plan for Estonia: Summary and conclusions.* Tallinn, Ministry of Industry and Energetics, Republic of Estonia

Misiunas, R. & Taagepera, R. (1993). *The Baltic States: Years of dependence 1940-1990*. London, Hurst & Company.

Missfeldt, Fanny (1999). Game-theoretic modelling of transboundary pollution. *Journal of Economic Surveys* 3(13): 287-321.

Molander, P. (1984). *Fångarnas dilemma - En översikt*. Stockholm, Försvarets Forskningsanstalt (FOA).

Monroe, K. R. (1997). "Review of Pathologies of rational choice theory: A critique of application in political science by Donald P. Green & Ian Shapiro." Political Theory 25(2): 289-95.

Morgenthau, H. (1973). *Politics among nations*. New York, Alfred A. Knopf.

Morrow, James D. (1994). *Game theory for political scientists*. Princeton, New Jersey, Princeton University Books.

Nash, P. (1981). *The Social Appraisal of Projects.* London, The McMillan Press Ltd.

NEFCO (1996). Nordic Environment Finance Corporation. Helsinki, NEFCO.

(1998). *Årsredovisning 1998*. Helsingfors, NEFCO.

Niou, E. M. S., Ordeshook, P. C., and Rose, G. F. *The balance of Power: Stability in international systems.* New York: Cambridge University Press.

Norborg, L.-A. (1988). Sveriges historia under 1800- och 1900-talen: Svensk samhällsutveckling 1809-1986. Stockholm, Norstedts Tryckeri.

Nordic Baltic Investment Committee (1995). *Baltic Investment Programme - evaluation report* (October 1995). Nordic Council of Ministers, Copenhagen.

Nordic Council of Ministers (1996). Samarbete mellan Baltikum och nordiska länderna för reduktion av koldioxidemissioner. Köpenhamn.

North, D. C. (1981). *Structure and change in economic history.* New York, Norton.

- 1990. Institutions, institutional change, and economic performance. Cambridge, Cambridge University Press.

Nozick, R. (1995). *The nature of rationality.* Princeton, Princeton University Press.

OECD (1996). *Regional integration and transition economies: The case of the Baltic rim.* Paris, Organisation for Economic Cooperation and Development, the Centre for cooperation with the economies in transition.

(1997). Environmental Data Compendium. OECD.

Oja, Tõnu (1993). System analysis approach in environmental planning and policy: Designing sustainable society (a case study – Estonia's own way). In *Energy, environment and natural resources management in the Baltic Sea region – 4th international conference on system analysis.* Report 1993: 653 p. 189-96. Copenhagen, the Nordic Council of Ministers.

Ojava, Pavel (1994). Oil terminals and ports. In Proceedings of the international conference on oil terminals, shipping and off-shore activities in the Eastern Baltic: Present situation, future prospects, economic and technological appraisal and environmental risk, page 25-32. 11-14 May, 1994. Tallinn, Stockholm Environment Institute-Tallinn.

Olson, Mancur (1965). *The logic of collective action.* Cambridge, Harvard University Press.

Opp, K.-D. (1999). "Contending conceptions of the theory of rational action." *Journal of Theoretical Politics* 11(2): 171-202.

Ordeshook, P. (1986). *Game theory and political theory. An introduction.* Cambridge, Cambridge University Press.

Ostrom, E. (1990). Governing the commons; the evolution of institutions for collective action. New York, Cambridge University Press.

- 1994. "Constituting social capital and collective action." *Journal of Theoretical Politics* 6(4): 527-62.

- 1998. "A behavioral approach to the rational choice theory of collective action." *American Political Science Review* 92(1): 1-22.

Ostrom, Gardner & Walker (1994). *Rules, games and common-pool resources.* Ann Arbor, The University of Michigan Press.

Ostrom, E & Keohane, R (1994). "Theoretical Puzzles." *Journal of Theoretical Politics* 6(4): 403-28.

Oye, M. (1994). "Self-interest and environmental management." *Journal of Theoretical Politics* 6(4): 593-624.

Palmlund, T. (1986). Altruism and other motives: Swedish development aid and foreign policy. In *Swedish development aid in perspective: Policies, problems & results since 1952.* P. Frühling. Värnamo, Fälths Tryckeri AB: 109-128.

Pedersen, J. D. (1994). *The Baltic Sea: A hot-spot in an unfit plot.* Roskilde, Dept. of Environment, Technology and Social studies.

Persson, Torsten & Tabellini, Guido. (1993). *Designing institutions for monetary stability.* Carnegie-Rochester Conference Series on Public Policy, Elsevier Science Publishers BV.

- 1996. Separation of powers and accountability: towards a formal approach to comparative politics. Stockholm, Institute for International Economic Studies.

Petersens, F. A. (1996). *BITS sexton år i biståndet 1979-1995*. Stockholm, BITS.

Priiman, Reet (1993). About natural bioadsorption water treatment. In Energy, environment and natural resources management in the Baltic Sea region – 4th international conference on system analysis. Report 1993: 653 p. 321-24. Copenhagen, the Nordic Council of Ministers.

Putnam, R. (1988). "Diplomacy and domestic politics: the logic of two-level games." *International Organization* 42(3): 429-60.

Radetzki, M. (1994). Sweden's assistance to Poland: An overview evaluation. Stockholm, SASDA.

Rasmusen, E. (1997). *Games and information: An introduction to game theory*. Oxford, Blackwell Publishers Ltd.

Ringius, L., Torvanger, A. & Mezears, E. (1996). Climate policy, burden sharing and the Nordic countries - present state of analysis and need for further analysis. Copenhagen, CICERO.

Ringius, L., Holm, J. & Klemmensen, B. (1996). Denmark's environmental aid to Eastern Europe: Present and future. In Löfstedt, R. E. & Sjöstedt, G (ed), *Environmental aid programmes to Eastern Europe: Area studies and theoretical applications*. Brookfield USA, Aldershot.

Ringius, L. (1996). The environmental action plan approach: A milestone in pollution control in the Baltic Sea. In Hjorth, R., *Baltic environmental cooperation. A regime in transition*. Motala, Kanaltryckeriet.

Rittberger, V., Ed. (1993). *Regime theory and international relations*. Oxford, Clarendon Press.

Roose, Antti (1991). *Estonia built on oil shale*. Rakvere, Virumaa Foundation.

Ross, Juhan, Kirt, Enno, Koppel, Andres & Kull, Kalevi (1993). Swedish-Estonian forestry project. In Energy, environment and natural resources management in the Baltic Sea region – 4th international conference on system analysis. Report 1993: 653 p. 103-104. Copenhagen, the Nordic Council of Ministers.

Salay, J., Fenhann, J, Jaanimägi, K. & Kristoferson, L. (1993). "Energy and the environment in the Baltic States". *Annual Review of Energy and the Environment*, 6: 169-216.

Scharpf, F. W. (1997). Games real actors play: Actor-centered institutionalism in policy research. Oxford, Westview Press.

Schelling, T. C. (1960). *The strategy of conflict*. Cambridge.

Schipper, Lee, Martinot, Eric, Khrushch, Marta, Salay, Jürgen, Sheinbaum, Claudia & Vorsatz, Diana (1994). *Structure and efficiency of energy use in a reforming country: The case of Estonia*. Stockholm, Stockholm Environment Institute.

Schrøder, Niels & Kjærgård, Bente (1993). Environmental safe handling of the residues from energy production, iron recovery, waste incineration and military waste. In *Energy, environment and natural resources management in the Baltic Sea region – 4th international conference on system analysis*. Report 1993: 653 p. 61-72. Copenhagen, the Nordic Council of Ministers.

Schweers, Karen & Levi, Maragaret (1990). *The limits of rationality*. Chicago, University of Chicago Press.

Selin, Henrik (2000). Towards international chemical safety: Social learning and the formation of the regime on persistent organic pollutants under the Convention on Long-Range Transboundary Air Pollution (dissertation no. 211). Linköping, Linköping University.

Sen, A. (1977). "Rational fools: A critique of the behavioral foundations of economic theory." *Philosophy and Public Affairs* (6): 317-44.

Shapiro, I. (1994). Pathologies of rational choice theory: A critique of applications in political science. New Haven, Yale University Press.

Shepsle, K. A. & Weingast, B. R. (1984). Legislative politics and budget outcomes. *Federal Budget Policy in the 1980s*. G. M. J. Palmer. Washington D. C., Urban Institute Press: 343-67.

- 1996. Statistical political philosophy and positive political theory. In *The rational choice controversy*. J. Friedman. New Haven and London, Yale University Press.

Shepsle, Kenneth A & Bonchek, Stephen (1997). *Analyzing politics*. New York, W. W. Norton.

Sida (1996a). *Sida-Östs insatser på energiområdet i Central- och Östeuropa*. Förslag till inriktning och former. Stockholm, Sida.

- 1996b. *Sidas stöd till Östersjöprogrammet*. Stockholm, Avdelningen för Central och Östeuropa.

- 1996c. *Insatser i Central- och Östeuropa*. Stockholm, Avdelningen för Central och Östeuropa.

- 1997. *Environment and energy projects*. Stockholm, Dep. for Central and Eastern Europe.

Sjöstedt, Gunnar (1996). Environmental aid negotiations as a process of collective learning. In Löfstedt, R. E. & Sjöstedt, G (ed), *Environmental aid programmes to Eastern Europe: Area studies and theoretical applications*. Brookfield USA, Aldershot.

Slote, M. A. (1989). *Beyond optimizing: a study of rational choice*. Cambridge, Mass.: London, Harvard University Press.

Smith, G. (1996 (1994)). The Baltic States: The national self-determination of Estonia, Latvia and Lithuania. Hampshire, Macmillan Press Ltd.

Snidal (1991). "Relative Gains and Patterns of International Cooperation." *American Political Science Review* 85.

SNV (1996). Lead party report on Combined municipal and industrial discharges. Stockholm, Statens Naturvårdsverk.

- 2000a. Environmental monitoring for a healthier Baltic. *Enviro Report* No. 3.

- 2000b. The Baltic – an endangered sea. *Enviro Report* No. 3.

- 2000c. Stresses on sensitive Baltic gradually. *Enviro Report* No. 3.

SOU (1990: 88). Sveriges internationella miljösamarbete: Nya mål och nya möjligheter. Stockholm, Miljödepartementet.

- 1991: 48. *Bistånd genom internationella organisationer*. Stockholm, Utrikesdepartementet.

- 1994: 102. Analys och utvärdering av bistånd, betänkande av kommittén för analys av utvecklingssamarbete. Stockholm, UD.

Stålnacke, P. (1996). *Nutrient loads to the Baltic Sea* (dissertation no. 146). Linköping, Linköping University.

Steen, A. (1997). "The new elites in the Baltic States: Recirculation and change." *Scandinavian Political Studies* 20(1): 91-112.

Steinrücke, Jürgen (1993). The possible use of wind energy on the Estonian coast and other ways of reducing of CO_2 in Estonia. In *Energy, environment and natural resources management in the Baltic Sea region – 4th international conference on system analysis*. Report 1993: 653 p. 139-146. Copenhagen, the Nordic Council of Ministers.

Stiegler George J. & Becker Gary S. (1977). "De gustibus non est disputandum." *American Economic Review* 67: 76 -90.

Stockholm Environment Institute (1994). Beauty and the East - An evaluation of the Swedish environmental assistance to Eastern Europe. Stockholm, SASDA.

Stone, R. W. (1999, February 16-20). *The use and abuse of game theory in international relations: the theory of moves*. 40th Annual Convention of the International Studies Association, Washington D.C.

Strange, S. (1982). "Cave! hic dragones: a critique of regime analysis." *International Organization* 36(2): 337-56.

Szrubka, Wojciech (forthcoming). *The legitimation of bureaucratic power*. Dissertation. Linköping, Linköping University.

Tammemäe, Olavi (1994). Environmental standards of oil terminals, ports and transport, environmental risk and technological appraisal in Estonia. In *Proceedings of the international conference on oil terminals, shipping and off-shore activities in the Eastern Baltic: Present situation, future prospects, economic and technological appraisal and environmental risk*, page 15-24. 11-14 May, 1994. Tallinn, Stockholm Environment Institute-Tallinn.

Tarand, Andres (1996). Environmental aid to Estonia. In Löfstedt, R. E. & Sjöstedt, G (ed), *Environmental aid programmes to Eastern Europe: Area studies and theoretical applications*. Brookfield USA, Aldershot.

Taylor, M. (1976). *Anarchy and cooperation*. London, Wiley.

Tomik, Arvi, Samlan, Ülo & Kaljuvee, Erik (1993). Development of oil shale mining and the state of the land surface. In Energy, environment and natural resources management in the Baltic Sea region – 4th international conference on system analysis. Report 1993: 653 p. 105-114. Copenhagen, the Nordic Council of Ministers.

Tomson, Teolan & Pajumets, Eha (1993). Utilization of solar thermal energy under the conditions of Estonia. In Energy, environment and natural resources management in the Baltic Sea region – 4th international conference on system analysis. Report 1993: 653 p. 177-180. Copenhagen, the Nordic Council of Ministers.

Tsebelis, G. (1996). *Nested games: Rational choice in comparative politics*. Los Angeles, University of California Press.

Tähtinen, M. & Nurste, H. (1992). *Energy use and emission scenarios to the year 2000 for Estonia.* Espoo, Tech. Res. Cent. Finland.

Udehn, L. (1996). The limits of Public choice; a sociological critique of the economic theory of politics. London, Routledge.

Underdal, A. (1991). The outcomes of negotiations. In *International negotiations: Analysis, approaches, issues*. V. A. Kremenyuk. San Francisco, Jossey-Bass Publishers: 100-16.

UNDP, United Nations Development Programme (1996a). UNDP country cooperation frameworks and related matters. First country cooperation framework for Latvia (1997-99). December 1996, UNDP.

- 1996b. *Lithuania Human Development Report*. Vilnius, UNDP Lithuania.

- 1997. Lithuanian Human Development report. Vilnius, UNDP Lithuania.

Utrikesdepartementet (1994a). *Is Swedish aid rational: A critical study of Swedish aid policy in the period 1968-1993*. Stockholm, Secretariat for analysis of Swedish development assistance.

- 1994b. Hållbart bistånd - det svenska biståndet efter UNCED (Ds 1994: 132 volym 1). Stockholm.

- 1996. Databas över miljösamarbete med CEE (19890101-19960701).

Uusi-Kämppä, Jaana & Yläranta, Toivo (1993). Buffer strips for sediment and nutrient removal from agricultural surface runoff. In *Energy, environment and natural resources management in the Baltic Sea region – 4th international conference on system analysis.* Report 1993: 653 p. 469-78. Copenhagen, the Nordic Council of Ministers.

Vaahtoranta, S. (1994). "The interest-based explanation of international environmental policy." *International Organization* 48(1): 77-105.

Walt, Stephen, M. "Rigor or rigor mortis? Rational choice and security studies". *International Security* 23(4): 5-48.

Waltz, K. (1979). *Theory of International Politics*. New York, McGraw Hill, inc.

VanDeever, Stacy D. & Dabelko, Geoffrey D. (1999). "Redefining security around the Baltic: Environmental issues in regional context." In *Global Governance* 5: 221-49.

Weale, A. (1992). *The new politics of pollution*. Manchester, Manchester University Press.

Vebra, Evaldas (1996). Country report: Lithuania. In *Approximation of European Union Environmental legislation.* The Regional Environmental Center for Central and Eastern Europe (REC). Budapest, REC.

Weyland, K. (1996). "Risk taking in Latin American economic reconstructing: Lessons from prospect theory." *International Studies Quarterly* (40): 185-208.

Whalley, J. (1991). "The interface between environmental and trade policies." *The Economic Journal* 101: 180-189.

Vihalemm, Peter (2000). "Changing Baltic space: Estonia and its neighbors." *Journal of Baltic Studies* XXX(3): 250-69.

Vogel, Michael P. (ed). (1999). Environmental Kuznets Curves: A study on the economic theory and political economy of environmental quality improvements in the course of economic growth. Lecture notes in economics and mathematical systems no. 469. Berlin, Springer .

Vought, Lena B. M., Lacoursiere, Jean O., Pedersen, Carsten L. & Dahl, Jonas (1993). Buffer-strips in agricultural landscape. In *Energy, environment and natural resources management in the Baltic Sea region – 4th international conference on system analysis.* Report 1993: 653 p. 393-94. Copenhagen, the Nordic Council of Ministers.

Yamaguchi, Kazuo (2000). "Subjective rationality of initiators and threshold-theoretical behavior of followers in collective action: theoretical insight from non-expected utility models." *Rationality and Society,* 12(2): 285-226.

Young, Oran R. (1982). *Resource regimes: natural resources and social institutions.* Berkeley, University of California Press.

- 1991. International Cooperation: Building Regimes for Natural Resources and the Environment. London, Cornell University Press.

Young, Oran R. & Osherenko, Gail, Ed. (1993). *Polar politics.* Ithaca, N.Y., Cornell University Press.

Zagare, F. (1987). *The dynamics of deterrence.* Chicago, University of Chicago.

Zermelo, E. (1913). *Über eine Anwendung der Mengenlehre auf die theorie des schachspiels.* Fifth International Congress of Mathematicians.

MIX
Papier aus verantwortungsvollen Quellen
Paper from responsible sources
FSC® C105338

If you have any concerns about our products,
you can contact us on
ProductSafety@springernature.com

In case Publisher is established outside the EU,
the EU authorized representative is:
Springer Nature Customer Service Center GmbH
Europaplatz 3, 69115 Heidelberg, Germany

Printed by Libri Plureos GmbH
in Hamburg, Germany